POULTRY BEHAVIOUR AND WELFARE

Michael C. Appleby, PhD

The Humane Society of the United States
Washington, DC
USA
Formerly University of Edinburgh, UK

Joy A. Mench, DPhil

Department of Animal Science
University of California
Davis, California
USA

and

Barry O. Hughes, PhD

Roslin Institute
Edinburgh
UK

CABI Publishing

CABI Publishing is a division of CAB International

CABI Publishing
CAB International
Wallingford
Oxfordshire OX10 8DE
UK
Tel: +44 (0)1491 832111
Fax: +44 (0)1491 833508
E-mail: cabi@cabi.org
Website: www.cabi-publishing.org

CABI Publishing
875 Massachusetts Avenue
7th Floor
Cambridge, MA 02139
USA
Tel: +1 617 395 4056
Fax: +1 617 354 6875
E-mail: cabi-nao@cabi.org

A catalogue record for this book is available from the British Library,
London, UK.

Library of Congress Cataloging-in-Publication Data
Appleby, Michael C.
  Poultry behaviour and welfare / Michael C. Appleby, Joy A. Mench, and
Barry O. Hughes.
     p. cm.
Includes bibliographical references and index.
  ISBN 0-85199-667-1 (alk. paper)
  1. Poultry. 2. Poultry--Behavior. I. Mench, Joy A. II. Hughes, Barry O.
III. Title.
SF487 .A64 2004
636.5--dc22

                          2003025184

ISBN 0 85199 667 1

Typeset in 10/12pt Baskerville by Columns Design Ltd, Reading
Printed and bound in the UK by Cromwell Press, Trowbridge

b
3.6.05

POULTRY BEHAVIOUR AND WELFARE

Contents

Preface

To a much greater extent than most people would imagine, humans depend on poultry for their food. There are about twice as many chickens in the world as humans, and most, with all the other poultry species, are kept for meat and eggs. Yet the contributions of poultry to human society remain widely ignored or unknown, their variety unappreciated and the complexity of their behaviour overlooked or belittled. This volume has three overlapping themes. First, that chickens, turkeys, ducks, geese and other poultry are fascinating representatives of the world of birds, with biology and behaviour that are novel to those who are mostly familiar with mammals. Secondly, that knowledge of their biology and behaviour is essential to proper management of commercial poultry: witness the facts that success in management is considerably affected by experience, and that decisions made on a technical rather than a biological basis are often disastrous. Thirdly, that such knowledge fosters an appropriate concern for the birds themselves, a concern that mirrors increasing public interest in the welfare of farm animals. To have a sustainable future, agriculture must take into account the needs of our animals and our environment as well as ourselves.

This volume draws on some material from the previous *Poultry Production Systems: Behaviour, Management and Welfare* (Appleby *et al.*, 1992), revised and updated. However, the issues concerned have moved on considerably in the last 10 years and there is much new coverage, particularly in the chapters on welfare, politics and economics. There is still more published information available on chickens than on other poultry, but studies on the latter are increasing and are included as comprehensively as possible. This continues to be an active area of research and more than one-third of the references listed have been published since the previous book.

We are very grateful to many friends and colleagues for support during the production of this volume. We particularly wish to thank Francine Bradley, Ben Mather and Anna Olsson who each read the whole manuscript and made many helpful comments. M.C.A. would like to dedicate his contribution to his mother, Josephine M. Appleby, a biologist and teacher whose lifelong interest in and concern for animals has inspired many others as well as himself. J.A.M. dedicates her input to her parents, Walter and Lucille Mench, for a lifetime of support and encouragement. B.O.H. wishes to dedicate his contribution to his wife, Helen Hughes, for her encouragement, love and patience.

1   Origins

1.1 Summary

- Progenitors of poultry came from few taxonomic groups and shared features suitable for domestication, including behaviour. For example, they show sexual promiscuity and flexible dietary requirements. Despite centuries of selection for conformation and plumage and decades of selection for eggs and meat, most behaviour of wild relatives is also shown by modern poultry.
- Domestic fowl were domesticated from red jungle fowl over 8000 years ago, and the basis of modern breeds existed by Roman times. Modern commercial hybrid chickens have been selected for maximum egg or meat output from minimum food intake. New genetic techniques will increase the tendency to produce lines with very specific characteristics.
- Turkeys were domesticated in the Americas over 2000 years ago. Most commercial birds are now white feathered, and very heavy. Fertility problems with natural mating mean artificial insemination is commercially routine.
- Japanese quail were first domesticated about 1000 years ago, but systematic selection for egg and meat production began about 100 years ago. Bobwhite quail are more common in the Americas, where they are still used as game birds as well as for meat and sometimes eggs.
- Guinea fowl came from West Africa and pheasants from Central Asia. Guinea fowl are grown mainly for meat, while pheasants are still mostly game birds, but some are now bred for meat.
- Domestic ducks derived from the mallard. Most breeds are kept for meat but some are prolific egg layers. Muscovy ducks are from Central and South America, and have less fat than domestic ducks. Domestic geese are primarily descended from the greylag in Asia, and their husbandry was well developed in Roman times.
- Domestic pigeons, derived from the rock dove, are now a minority interest in most countries. In contrast, farming of ostriches (and to a lesser extent rheas and emus) has increased in recent decades.

1.2 Domestication and Behaviour

All domestic poultry other than ratites come from three Orders, the Galliformes (including fowl), the Anseriformes (ducks and geese) and the Columbiformes (pigeons). This emphasizes the fact that their progenitors must have shared biological features, including aspects of behaviour, which predisposed them for domestication.

Birds were first domesticated for their behaviour – to be used for cockfighting – and behaviour is important in many other aspects of domestication. Hale (1962) has pointed out that species, both mammals and birds, which have been successfully domesticated share a number of common features. They form relatively large groups and have a hierarchical structure with males affiliated to female groups. They show promiscuous mating, males are dominant over females and sexual signals are behavioural, rather than by colour markings or morphological structures. These features allow the animals to be easily managed in large numbers, while the maintenance of hierarchies through social dominance reduces the danger of injury caused by constant fighting. Promiscuous sexual behaviour allows any male to be mated with any female. Alterations to markings or structures often occur during selection programmes, so their irrelevance to successful mating is very helpful. Parent–young interactions are important; favourable characteristics include a critical period of bond development such as imprinting, the acceptance by parents of other young soon after hatching and precocial development of the young. This allows animals to be readily tamed, because they can bond to humans rather than to their parents, and to be reared by surrogate parents if required. Precocial development minimizes the length of time during which the young require specialized care and development. Animals that show favourable responses to humans, such as a short flight distance and little disturbance due to human activities, are also well suited to domestication. Other behavioural characteristics that are helpful include flexible dietary requirements, especially the ability to forage, which favours seed or grass eaters (Sossinka, 1982), adaptability to a wide range of climatic and environmental conditions, and limited agility. Domesticated birds show many of these features.

We shall outline the origins and domestication of the main species of poultry, starting with the Galliformes.

1.3 Domestic Fowl

The progenitor of the domestic fowl was the red jungle fowl (*Gallus gallus*), modern forms of which are found in Central and South India (*Gallus gallus sonnerati*), East India (*G. g. murghi*), Burma and Malaysia (*G. g. spadiceus*) and Thailand and Cambodia (*G. g. gallus*). It is a smaller bird than most domestic varieties – an adult female weighs about 800 g – and it is a tropical species. Along the Himalayan foothills, its range is bounded by the 10°C isotherm, and it is typically confined to forested areas and to thick vegetation. However, the jungle fowl's ability to adapt to a broad range of environments, together with its potential genetic variability, subsequently helped the domestic fowl to become widely distributed throughout the world. It is believed that the fowl was first domesticated in South-east Asia, probably well over 8000 years ago (Yamada, 1988). The Latin nomenclature of the domestic fowl has been a source of controversy, and for a long period it was termed *G. domesticus*. It is now accepted that it is not a separate species but should be regarded as a subspecies of the jungle fowl and thus be called *G. g. domesticus*. Until recently, it was thought that the fowl reached Europe via India and the Middle East, because of the evidence of domestication in the Indus Valley around 2000 BCE (before the common era) (Sewell and Guha, 1931; Zeuner, 1963) and the fact that it was known to be present in the area around Babylon in 2400 BCE and in Egypt in 1400 BCE. Early Sumerian texts contain the word for cock, and there is a clear representation of a galliform cockerel outside Tutankhamen's tomb (Coltherd, 1966). However, recent archaeological research, based primarily on the existence of chicken bones in deposits of known age, has shown that after domestication, domestic fowl were first taken north (Fig. 1.1); they were established in China by 6000 BCE (West and Zhou, 1989). From there, it is believed that they were taken across the Russian steppes and there is evidence of their presence in Turkey and in Eastern Europe (Romania and Greece) during the later Stone Age (3000 BCE). By 1200 BCE, they had reached Spain, and they were to be found in north-west Europe by about 500 BCE (West and Zhou, 1989). The earliest records in Britain date from 100 BCE (Brown, 1929), and domestic fowl were introduced to North America about 1550 CE (Yamada, 1988).

During the earlier stages of domestication, the fowl was probably valued mainly as a sacrificial or religious bird, or for cockfighting. It was the Romans who developed its potential as an agricultural animal, creating specialized breeds (Thomson, 1964), including very productive layers, and forming a complex poultry industry, which paid close attention to rearing, housing, disease control, costing and marketing (Wood-Gush, 1959a). Pliny wrote that in Roman times there were birds laying an egg every day (Wood-Gush, 1971). The Romans knew about force feeding, hybrid vigour, caponizing and even sperm competition (Crawford, 1990). With the decline of the Roman Empire, the industry collapsed, fowls became little more than farmyard scavengers and poultry keeping did not resume on a large scale until the 19th century.

Wood-Gush (1959a) identifies several breeds in Roman times – two heavy fighting breeds, two dual-purpose ones, native Roman breeds and an especially prolific variety from Adria. Subsequently, little systematic selection was practised for many centuries, with the exception of birds for cockfighting. The appreciable levels of bird-to-bird aggression, which can pose a problem in some modern poultry-keeping systems, may be a legacy of this approach. It was only in the 19th century that the situation changed, with an explosion of poultry breeding. Six breeds are mentioned as existing in England around 1810: the Game, the White or English, the Black or Poland, the Darkling, the Large or Strakeberg and the Malay (Wood-Gush, 1959a). In the next 50 years, formal poultry shows were organized, many novel breeds were created and numerous Breed Societies were founded. Modern breeds are derived from two main types: Asiatic (e.g. Brahma, Cochin, Pekin and Malay) and Mediterranean (e.g. Ancona, Andalusian, Leghorn and Minorca). Other breeds were developed by crossing and selecting, their names often indicating their geographical origin. The Scots Grey, for example, originated in Scotland over 200

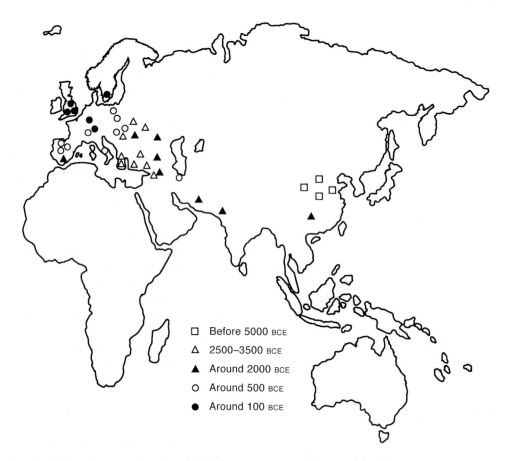

Fig. 1.1. Archaeological evidence shows that the fowl was first domesticated in South-east Asia and subsequently spread north and west through China. The symbols show the earliest dates of evidence of domesticated fowls at the various locations (after West and Zhou, 1989).

years ago, while the Sussex was listed in an English poultry show in 1845 and its make-up includes contributions from the Brahma, Cochin and Dorking. Similarly, in America, Plymouth Rocks, Wyandottes, Rhode Island Reds and other breeds were being developed towards the end of the 19th century. In addition to the large fowls, there are a number of bantams, which are miniature versions of the large breeds, often resembling them closely in conformation and plumage, but with a much smaller body size. There are 58 large fowl and 11 bantam breeds listed in the *British Poultry Standards Handbook* (May and Hawksworth, 1982), and 51 large fowl breeds and 62 breeds of bantams listed in the *American Standard of Perfection* (American Poultry Association, 1989). Additional breeds are recognized in other countries.

The last 30 years have seen the rise of the commercial hybrid; these are strains and lines rather than breeds. The hybrids are of two main kinds, egg-laying and meat. Both have been selected for 'efficiency' – maximum output for minimum food intake. The egg-laying types can be subdivided further into two classes. Light hybrids, primarily from White Leghorn, with mature female body weights around

1.5 kg and white egg shells, are favoured in continental Europe and the USA, while medium hybrids, derived primarily from Rhode Island Reds, with body weights of about 2.0 kg, lay brown eggs and are especially popular in the UK (Fig. 1.2). Both types have been exposed to intensive selection for egg number and quality, early maturity and food conversion efficiency. This means that body size and food intake have progressively shrunk, while egg number and size have increased. The meat-type hybrids originated from heavy breeds such as Cornish and White Plymouth Rock, and have been selected for growth rate, meat yield and proportion of white meat, as well as for high food conversion efficiency. This has resulted in a fast-growing, large, inactive bird and a carcass with a very high proportion of meat.

The globalization of the industry has resulted in amalgamation and the survival of the fittest, and only a handful of breeder companies now produce all the commercial birds (Chapter 11). A key recent development is the increased emphasis on single genes rather than type breeding. One example is the development of dwarf strains of meat-type origin; the dwarf gene is expressed in the breeding female, which has a lower food intake, is relatively small and is easier to manage, but not in the offspring, which grow much faster. Another is the naked neck broiler: the gene inhibits feather growth on the head and neck, thus increasing the rate of body heat loss, which is important in tropical climates where heat stress is a serious problem. With the recent elucidation of the poultry genome, there is now the possibility of identifying genes responsible for specific traits, and this will make possible the production of lines with very specific characteristics.

1.4 Turkey

Wild turkeys originated in North America, with the fossil record showing that they date back at least 10 million years. When the first Europeans arrived about 500 years ago, two species were present (Brant, 1998). The Latin name of the turkey, *Meleagris gallapavo*, was given to it by Linnaeus and reflects his incorrect belief that it was closely related to the guinea fowl; the second species, which has not been domesticated, is the ocellated turkey *M. ocellata*. The turkey was domesticated in Mexico, probably over 2000 years ago, and had spread through Central America by 700 CE (Crawford, 1990). The early Spanish explorers brought turkeys back to Spain, and they were distributed quickly over Europe, reaching Germany by 1530 and England by 1525 (Thomson, 1964) or 1541 (Brant, 1998). The common name of the bird is also misleading, and probably arose because most of the exotic birds being brought into Europe in the 16th century were coming from the east through Turkey.

The domestic bird was then taken back into the eastern USA by European colonists and was present in Virginia by 1607. The tradition in the USA of eating turkey at Thanksgiving dates back to early settlers in Massachusetts celebrating a successful harvest in 1621, but several centuries were to pass before the turkey became an important source of meat. During the early years of domestication and breeding, most of the emphasis was on plumage, with a large number of different varieties being developed for show purposes. It was not until the 20th century that meat production and conformation became important traits for selection (Brant,

Fig. 1.2. Four important breeds of fowl from which most of our modern hybrids have been derived are shown. The first three are layers, the last is a meat breed from which broiler strains were selected. Above: (A) White Leghorn. (B) Rhode Island Red. Opposite: (C) Light Sussex. (D) Plymouth Rock (May and Hawksworth, 1982). Male (left) and female (right) are shown for each breed.

1998). An English breeder called Throssell selected a heavyweight line called the Sheffield Bronze for the Christmas trade and subsequently emigrated to British Columbia with his stock. Canadians, impressed by his birds, crossed them with local strains to create the Broad-breasted Bronze (Brant, 1998).

(C)

(D)

By the late 1930s, with mature males weighing 18 kg and females 9 kg, this strain selected for meat production was spreading throughout the world. At the same time, the industry's need for smaller but faster-growing birds led to strains such as the Beltsville White, with mature males of 7 kg and females of 4 kg. There was also selection against the dark plumage of the wild turkey, in order to improve the appearance of the carcass by removing melanin from the feather follicles, and by the late 1960s most strains were white.

Selection for very heavy birds with substantial breast muscles resulted in fertility problems, as the conformation of the males made it increasingly difficult for them to

mate. The process would have been self-limiting had it not been for the development of the technique of artificial insemination, which is routinely used commercially.

1.5 Japanese Quail

The common quail (*Coturnix coturnix*) is widely spread through Europe, Africa and
Asia, but all domesticated birds have been derived from the Japanese quail, initially
classified as a subspecies but now regarded as a species in its own right (*Coturnix
japonica*). They were first domesticated about 1000 CE and kept for their song, but
systematic selection for improved egg and meat production began in Japan only
about 100 years ago. Widespread introduction to North America and Europe
occurred during the 1950s. The wild birds have an average weight of about 100 g,
females being slightly larger than males. The domesticated form is very similar in
plumage and general appearance but, because of selection for body weight, is now
typically 200–250 g. Egg number has also greatly increased and, like domestic fowl,
they are beginning to diverge into egg-laying and meat strains.

1.6 Bobwhite Quail

This species, *Colinus virginianus*, larger than *Coturnix* and from a different family
(Odontophorinae), is native to North and Central America. The male has a
distinctive call, from which the species' common name is derived. Bobwhite quail
are larger than Japanese quail, weighing about 450 g. There are more than 20
subspecies of bobwhite. The most common one kept in captivity is the eastern
bobwhite (*C. v. virginianus*), but commercial breeders may also breed masked, Texas,
plains and Florida bobwhite. The bobwhite is a popular game bird but, because of
the decline of wild populations due to habitat loss, it is now increasingly bred in
captivity. They are reared to be released into hunting preserves, for meat production
for gourmet food outlets and sometimes for egg production. Although bobwhite
could be considered to be only semi-domesticated, there has been increasing
selection pressure recently and the resultant development of strains with desirable
marketing characteristics, such as the Wisconsin Jumbo and the Indiana Giant bred
for large body size.

1.7 Guinea Fowl

The guinea fowl (*Numida meleagris*) has probably been domesticated for about 5000
years. The wild species is distributed across almost the whole of Africa south of the
Sahara and is found over a wide range of terrain, though it is most common on
savannah or in scrubland. Independent domestication probably occurred on a
number of separate occasions, because of the wild birds' tendency to associate with
human settlements and utilize resources such as food and water. It is probable,

though, that all modern domesticated birds are derived from the West African subspecies *N. m. galatea*. Guinea fowl in ancient Egypt were maintained in garden aviaries by wealthy nobles as an attractive feature. The first historical reference to them is found in an Egyptian mural dating from about 2400 BCE, and excellent representations appear on a Greek urn of the 6th century BCE. Well known in Rome by 30 BCE, when both eggs and meat were regarded as delicacies (Belshaw, 1985), they were re-introduced to Europe by the Portuguese exploring West Africa in the late 16th century. Currently, guinea fowl are produced commercially in some countries, especially for meat.

1.8 Pheasant

The wild form of the ring-necked pheasant, *Phasianus colchicus*, extends across Central Asia from the Black Sea to Manchuria. It was well known to the ancient Greeks and was probably distributed across Europe by the Romans. There were well-established stocks in France and England around 1000 CE, but additional introductions were made from Asian birds, beginning in the 18th century (Crawford, 1990). For many years, they were only semi-domesticated and this was a deliberate strategy to reduce tameness and thus make them more suitable as game birds, reared for shooting. Recently, however, some have begun to be kept in closer confinement and bred for meat production.

1.9 Domestic Duck

Of the Anseriformes, two species of ducks have been domesticated. That known as the domestic duck was derived from the mallard (*Anas platyrhynchos*), a Palaearctic species widely distributed over North America and Eurasia. All modern breeds appear to have been derived from the Eurasian subspecies *A. p. platyrhynchos*, which was first domesticated in South-east Asia or China 4000 years ago or even earlier. There is, however, speculation that it was also domesticated independently in the Middle East by the Sumerians around 2000 BCE and then again in Europe during the Middle Ages (Crawford, 1990). It was farmed by the Romans, who fattened them on net-covered ponds (Collias, 1962) but, because they are so easy to tame, these may have been semi-wild mallard rather than fully domesticated ducks. The mallard has given rise to a large number of different breeds: 18 are recognized in the UK (May and Hawksworth, 1982) and 14 in the USA (American Poultry Association, 1989). Selection for a number of traits has resulted in a wide variety of plumage and body conformations. The mallard weighs around 1.1 kg, but domestic duck breeds vary in size from the Appleyard Bantam at 700 g to the Aylesbury at 4.6 kg, with some commercial hybrids weighing up to 8 kg. Most, such as the Pekin from China, are kept for meat production, but some, such as the Khaki Campbell and Indian Runner, are extremely prolific egg layers, the latter being recorded as laying 363 eggs in 365 days (Jull, 1938).

1.10 Muscovy

The wild form of the Muscovy (*Cairina moschata*) is a forest duck indigenous to Central and South America, and had been domesticated before the arrival of Columbus. Its name is a mystery, but has no link to Moscow, except that it may have been brought to England around 1580 by the Muscovite Trading Company. It was imported to the Barbary Coast of Spain by the Spaniards around 1550, which explains the other name by which it is commonly known, the Barbary duck. The wild species has brownish-black plumage, with females weighing up to 1.5 kg and males around 3 kg. The domesticated form is similar but larger, has more white in its plumage and has a much more pronounced red facial caruncle. Being of tropical rather than Palaearctic origin, it has a much lower percentage of body fat than the domestic duck. Muscovy are hybridized with Pekin to produce Mulard ducks, which are raised mainly for the production of foie gras.

1.11 Goose

Domestic geese are primarily descended from the greylag (*Anser anser*). They may have undergone multiple domestication, but again they were first domesticated in China or South-east Asia, probably earlier than the duck. The Chinese goose may possibly be derived from a closely related species, *A. cygnoides* (Crawford, 1990). Both species are Palaearctic, the greylag being distributed from Europe to Manchuria, while *A. cygnoides* is found in Siberia. Domestic geese were well known in Europe by 700 BCE, as they are mentioned by the Greek poet Homer. The crested goose was valued by the Romans for guarding duties and reputedly saved the Capitol from the Gauls in 390 BCE by raising the alarm (Thomson, 1964). As with chickens, the Romans had a well-developed system for goose husbandry, with the birds being kept for meat, fat and feathers. Again it came to an end with the fall of the Roman Empire, with the birds reverting to farmyard scavenging. Eleven breeds are listed by May and Hawksworth (1982) and by the American Poultry Association (1989). The body weight of the wild species is around 3.5 kg; most domestic breeds are considerably larger.

1.12 Pigeon

Domestic pigeons are derived from the rock dove (*Columba livia*) in the Order Columbiformes and have been domesticated for at least 5000 years. The breeds and varieties of pigeon show considerable diversity: although some are similar to the wild form, others are very different. Some, such as the fan-tailed, have been bred for their plumage, and some for physical features, such as the pouter with its enormously enlarged crop region. Images of pigeons dating from 3100 BCE have been found in Egypt (Thomson, 1964). They were originally kept for eating; only later were they selected for their homing ability. In medieval England, they were unpopular with many people, for only the lord of the manor could maintain dovecotes, and the birds' food supply came from foraging on the peasants' crops. The young squabs were harvested for meat at the end of rearing, just before they were able to fly. Today they are mainly kept by aviculturalists and fanciers, and for racing.

1.13 Ratites

Ratites are large, flightless, walking birds with a flat rather than keeled sternum. Species include the rhea, emu and ostrich, but only the last has been truly domesticated. The ostrich (*Struthio camelus*) is the largest living bird (2.75 m tall and weighing up to 150 kg) and there are four subspecies, all resident in Africa (Deeming, 1999b). It was hunted for its meat, skin and feathers, and was first kept in captivity by the Egyptians, Greeks and Romans more than 2000 years ago (Shanaway, 1999). In more recent times, it was farmed in South Africa from around 1865 (Dingle, 1999), initially for its feathers, and later in Australia and the USA, until the market for plumes collapsed in 1914. Small-scale industry continued in South Africa, expanding from about 1950 onwards with the development of a market first for leather and then in 1985 for meat. Expansion was dramatic thereafter, with ostrich farms springing up in Australia, the USA and Europe. The modern, domesticated bird is a smaller, more docile hybrid of two subspecies (Dingle, 1999). However, its potential as a commercial species is limited by its low fertility, hatchability and chick survival (Deeming and Angel, 1996).

1.14 Breed Diversity

This long history of domestication and selection for traits that are important to humans has resulted in great biodiversity among the breeds of domestic poultry. With changing commercial and other pressures, there is great danger of many of these breeds and their unique genetic features being lost, so recently attempts have begun at an international level to survey, codify and preserve important populations (Weigend and Romanov, 2003).

2   Biology

2.1 Summary

- Though most poultry are poor fliers, their characteristic biological features are centred around adaptation to flight.
- The avian central nervous system is well developed, especially the cerebellum, which is concerned with control of flight, the optic lobes and other sensory areas, including those involved with hearing, taste, olfaction and the mechanoreceptors signalling touch, which are represented in the thalamus.
- The importance of the visual system is reflected by the large eyes, the visual acuity of many species and their tetrachromatic colour vision.
- Skeletal, muscular and respiratory systems are adapted to flight. For production purposes, this results in both advantages (such as the large breast muscles) and disadvantages (such as the light, aerated bones, which though strong are liable to break on impact).
- In laying strains of chickens, the female reproductive system, with its well-developed left ovary and oviduct, can produce up to 320 eggs a year. To achieve this output requires a very efficient digestive system, aided by the provision of a highly nutritious diet.
- Though in some respects more primitive than in mammals, the urinary system, coupled with reabsorption in the caeca, provides a very effective regulation of water balance. The immune system centres around the activity of lymphocytes, the B cells producing antibodies and the T cells destroying invading pathogens and eliminating damaged cells.
- The considerable knowledge now available on their biology should be taken into account when designing modern environments for poultry and also when selecting for production-oriented traits.

© M.C. Appleby, J.A. Mench and B.O. Hughes 2004. *Poultry Behaviour and Welfare*
(M.C. Appleby *et al.*)

2.2 Understanding Biology

The most characteristic feature of birds is their adaptation to flight; it is accordingly ironic that most types of poultry are poor or even non-fliers. Nevertheless, an understanding of avian biology is important when we consider the relationships between structure and function in poultry. There are several excellent books dealing with avian structure and function and with the biology and physiology of the domestic fowl (e.g. Bell and Freeman, 1971; King and McLelland, 1975, 1979–1985; Freeman, 1983–1984; Whittow, 2000).

2.3 Central Nervous System

Many of the anatomical features seen in reptilian brains have been preserved in the avian brain, but there is a massive increase in size and complexity, especially in the cerebral hemispheres, optic lobes and cerebellum (Fig. 2.1A). Even those birds with the least specialized degree of brain organization have brain masses 6–11 times greater than those found in reptiles of similar body size. Birds do not possess the extensive neocortex seen in mammals, and the majority of the cerebral hemisphere is composed of expanded striatal areas. The absence of neocortex does not, however, disadvantage the bird, as many of the functions of neocortical cells are assumed by striate neurones. With the complexities of flight and the importance of vision to birds, both the cerebellum and the optic lobes are correspondingly well developed. The importance of the beak in feeding and of the feet in movement is reflected by the distribution of mechanoreceptors in the basal nucleus of the thalamus; these are the parts of the body represented in the most detail in the central nervous system (Necker, 2000).

2.4 Vision

Vision is important to birds, and this is reflected in the fact that the avian eye is particularly large in relation to both the head and the brain: in the domestic fowl, the eyes together weigh about the same as the brain (Fig. 2.1B). Indeed, the eye of the ostrich is the largest of any land vertebrate. Pigeons and chickens have about 40% more retinal axons than humans, while birds of prey have better visual acuity than almost all other animals. The laterally placed eyes of herbivore and omnivore species such as poultry extend over a visual field of more than 300°, but they cover a much smaller binocular zone than predatory carnivorous species with frontally directed eyes (Fig. 2.2). The eye is protected not only by mobile upper and lower eyelids but also by a nictitating membrane originating in the medial canthus and moving laterally across the eyeball. It sweeps the lachrimal secretion across the cornea, removing any excess on its return movement. It is transparent in diurnal species and impairs vision so little that it is suggested that birds fly with it extended, to protect the cornea and prevent it drying out.

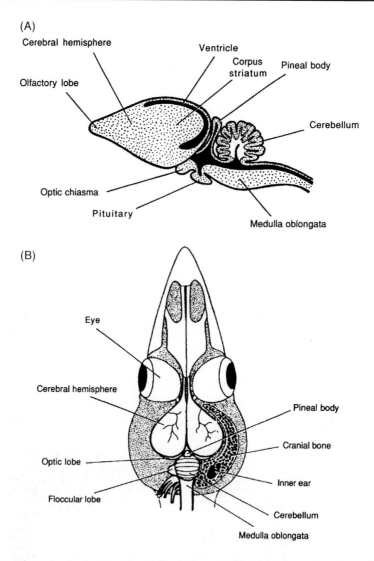

Fig. 2.1. The avian brain. (A) A saggital section of the brain to show the relatively large cerebral hemispheres, which coordinate the higher functions, and the cerebellum, concerned with the control of movement. (B) The fowl's skull from above with the top removed to show how most of the space is occupied by the brain and the eyes (after Ede, 1964).

Birds have good colour vision, diurnal species such as the fowl possessing more cones than rods. There is a central area, where the receptors are very closely packed to enhance optical resolution; it is circular in grain-eating birds but oval or band-shaped in water birds, probably to improve perception and recognition of objects on horizontal surfaces. Whereas humans have trichromatic vision, birds have four types of cone and can see further into the ultra-violet (UV) (Fig. 2.3). There is also a coloured oil droplet at the distal end of each cone; these both act as lenses focusing light on the photoreceptor and enhance colour discrimination and perception.

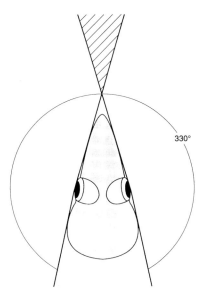

Fig. 2.2. Fowl's head from above showing how each eye has a very extensive field of view forwards, sideways and backwards, but that the area of binocular overlap (shown hatched) is relatively small.

Projecting from the rear surface of the retina is a comb-shaped, heavily vascularized and pigmented object called the pecten. Its precise function is uncertain, but it is larger in diurnal species and may reduce glare in bright light. Its other

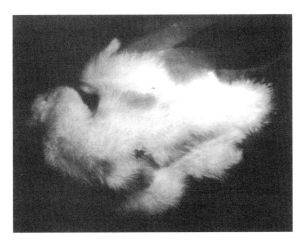

Fig. 2.3. Dorsal view of a 7-day-old turkey poult illuminated by ultra-violet (UV) light and photographed through a filter that removes frequencies below 415 nm. Under normal fluorescent light, the poult appeared uniformly yellowish-white, whereas, under UV, two types of marking were present: dark patches on the shoulder, while the primary feathers were a violet colour to the unaided eye (from Sherwin and Devereux, 1999, with permission from Taylor & Francis Ltd).

probable role is to increase diffusion of oxygen and nutrients into the aqueous humour because, in birds, to reduce light scatter and enhance visual resolution, the retina is avascular.

Experiments studying the visual perception of birds have shown that they respond to visual stimuli very much like human beings. If they are trained to choose the larger of two objects and then offered a choice between two that are objectively identical in size but which differ in orientation or presentation so that one appears subjectively larger to human beings, birds too select the subjectively larger one. Pigeons can learn visual concepts such as 'same versus different', 'animal' and even cartoon figures.

This high standard of visual acuity, sensitivity and perception is of little relevance to domestic birds kept in confined surroundings, such as hens in battery cages, but is much more important when they are housed under more extensive conditions. There it helps them to identify sources of food and water, suitable nest sites and locations to scratch, dust bathe and roost. It is also crucial for individual recognition, for identifying sign stimuli and thus for the maintenance of dominance hierarchies and the social order.

2.5 Hearing

Sensitive hearing and vocal communication are generally very important to birds. The frequency range to which birds are sensitive is about 15–10,000 Hz (Bremond, 1963): auditory threshold tests on the fowl have shown that, while they respond at frequencies from 250 to 8000 Hz, the range of most sensitive hearing lies between 3000 and 5000 Hz (Temple et al., 1984). Like vision, this sense is of key importance to precocious species. Chicks of the jungle fowl, for example, when only 1 day old respond to long-duration, high-frequency sounds (such as the overhead predator call) by avoidance behaviour such as squatting down or running away (Kruijt, 1964). The domestic fowl has a large repertoire of about 20 separate and distinguishable calls, each given in a separate and definable context (Fig. 2.4), which consist mainly of frequencies between 400 and 6000 Hz (Wood-Gush, 1971). Clearly, an acute and sensitive hearing ability is necessary to allow other individuals to distinguish these calls accurately so they can make appropriate responses.

The external opening to the ear is inconspicuous in birds and is surrounded by specialized feathers, which do not obstruct sound transmission; in some birds, such as owls, these are arranged so as to reflect sound into the canal. The anatomy of the ear is similar to that of mammals except that, in the middle ear, sound vibrations are carried from the tympanic membrane to the oval window by a single ossicle, the columella.

2.6 Taste

The fowl has a well-developed sense of taste. The taste buds, averaging about 350 in number with a maximum of 500, are located on the dorsal surface of the tongue

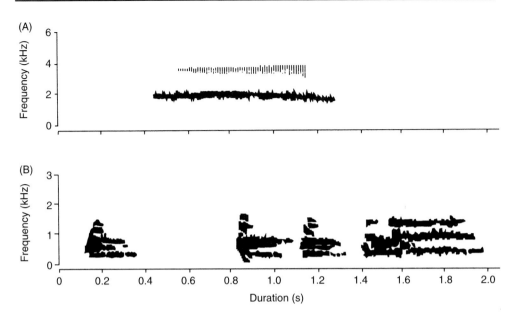

Fig. 2.4. Sound spectrographs of two of the domestic fowl's calls (after Wood-Gush, 1971). (A) The overhead predator call is of high pitch (penetrates well), long duration (commands attention) and gradual onset and termination (difficult to localize). (B) The ground predator call, in contrast, consists of a sequence of clucks.

and in crypts at the openings of the salivary glands in the roof and the floor of the oropharynx. Behavioural studies (Halpern, 1962; Gentle, 1975) have shown that birds' responses to flavours differ in certain respects from those shown by mammals such as the rat. Acid and bitter flavours are rejected by the domestic fowl, as they are by the rat, but whereas rats prefer weak and moderate salt solutions to pure water, fowls do not. Both reject strongly saline solutions. Sweet flavours, whether of natural or artificial origin, though selected strongly by rats, are generally not especially attractive to fowl, although they are to fruit- and nectar-feeding birds such as budgerigars or hummingbirds. It may be that fowls use visual and tactile senses for food selection much more than do mammals, the primary function of taste being to reject items that may be noxious.

Fowls are very sensitive to water temperature, are reluctant to drink warm water above ambient temperature and can discriminate differences as small as 3°C.

2.7 Olfaction

The nasal cavity of many species of birds has an olfactory epithelium that is structurally similar to that found in mammals, and it is possible to show positive neurophysiological reactions to various odours (Neuhaus, 1963). Operant conditioning, field experiments and observations of behavioural responses to olfactory cues also provide convincing evidence of a well-developed sense of smell in birds (Bang and Wenzel, 1985). There is a clear correlation between olfactory bulb size and the extent to which particular species use olfactory cues in nature (Bang and Wenzel, 1985). For some bird species, such as passerines, olfaction is not as important as it is for mammals, whereas for others it is essential for locating food: the albatross can detect meat at a distance of 20 miles (Miller, 1942). Kiwis, too, which have nostrils at the end of their upper mandibles, appear to use their sense of smell when probing in the ground for prey items. It is also known that, for migrating pigeons, odours carried on the wind can be an important navigational cue, while domestic fowls can be trained to respond selectively to particular scents, such as oil of citron (Jones and Gentle, 1985).

2.8 Cutaneous Sensitivity

The skin of the bird is well supplied with sensory receptors, especially those areas of the body not covered by feathers, such as the beak. In the beak, there are also concentrations of touch receptors grouped to form special beak tip organs, which allow the bird to make very fine tactile discriminations. Damage to the beak, of the kind imposed by beak trimming, will greatly impair birds' sensory abilities. There are three different types of touch receptors present: two which respond to a moving stimulus (Herbst and Grandry corpuscles) and one which responds to static pressure (slowly adapting mechanoreceptors). Environmental temperature is monitored by cold receptors that respond to cooling of the skin, and warm receptors that respond to heat. Noxious (unpleasant or painful) stimulation is detected by another group of receptors, the nociceptors, of which there are at least three types present in the domestic fowl and which respond to either severe pressure or major changes in temperature.

2.9 Skeletal and Muscular Systems

Birds differ from most mammals in two key respects: their bipedal stance and ability to fly. The lightweight bone structure of birds is an adaptation for flight, while part of the spinal column and pelvic girdle are fused to increase their rigidity and allow the entire body weight to be carried through the hind limbs (Ede, 1964). Development of massive pectoral muscles is another adaptation for flight. In largely flightless birds such as the domestic fowl, these muscles contain very little myoglobin and are therefore 'white'. Genetic selection for an increased yield of this desirable meat in both chickens and turkeys has been very successful in producing broiler-type birds, which grow four times faster with eight times more breast muscle mass than comparable layer types (Griffin and Goddard, 1994). However, this has also had undesirable consequences, some general, such as metabolic disorders (Scheele, 1997), and some reflected in musculoskeletal dysfunction, such as myopathy in broilers (Siller, 1985) and in turkeys (Wilson *et al.*, 1990). It appears that there may be a limit to muscle fibre hypertrophy, beyond which degeneration occurs, with breakdown of membrane integrity and biochemical changes such as release of creatine kinase into the circulation (Sandercock *et al.*, 2001). In turn, there will be adverse effects on welfare, including reduction of mobility.

Storage of calcium in medullary bone is an adaptation to egg laying. The store is built up before birds come into lay and is then mobilized for egg shell formation. During the ovulation–oviposition cycle, periods of intense medullary bone formation alternate with periods of severe depletion. Unless sufficient dietary calcium is available to replenish the medullary bone every day, then structural cortical bone is also mobilized and, over time, osteoporosis of the long bones occurs so that, by the end of the laying period, hens are at risk of bone fractures.

2.10 Female Reproductive System

The female avian reproductive system is unusual in that, although two gonads and oviducts begin developing in the embryo, those on the right side begin to regress quite early on while only the left ovary and oviduct continue to develop and become functional. The ovary grows especially rapidly with the onset of sexual maturity; in the domestic fowl, it reaches 60 g and occupies a mid-line position, overlapping the kidneys and lungs. It contains many thousands of oocytes, which develop sequentially into follicles. These grow very slowly up to about 2 mm diameter. A mechanism that is not fully understood then selects one follicle daily for rapid growth. It reaches the full size of 40 mm in about 8 days, when it is ready for ovulation and, together with the other six or seven large follicles of various sizes, gives the ovary the appearance of a bunch of grapes (Fig. 2.5). The primary oocytes of birds, the 'yolks' of eggs, are the largest cells in the animal kingdom, those of the domestic fowl each weighing about 20 g. At ovulation, the ovum is released into the abdominal cavity and is picked up by the fimbria (funnel) of the infundibulum (Fig. 2.6). It passes down the oviduct into the magnum region, which has a deep glandular layer secreting albumen. This forms the thick coating of egg white around the ovum. It then passes through the isthmus, where the egg membranes are formed. The membranes are of protein fibres and are semi-permeable, allowing water and ions to pass, but not albumen. The inner and outer membranes separate at the broad pole (large end) to form the air cell. The ovum then moves on into the shell gland (uterus). Here water is first transferred across the membranes, plumping the albumen, and calcium carbonate is then deposited to form the egg shell. The egg passes through the oviduct in about 25 h; about 20 of these are spent in the shell gland.

The shell itself has a number of important functions: it provides mechanical protection to the embryo, it prevents ingress of harmful organisms, it maintains a suitable medium for development, allowing diffusion of gases including water vapour, it acts as a calcium source and it is readily opened by the chick at the end of incubation. The calcified shell proper consists of the mammillary layer, which acts as the base for calcification; the palisade layer, which forms the major part of the shell and consists of layers of crystalline calcium carbonate interspersed with layers of organic protein matrix; and the cuticle. This last, though only 1–10 μm thick, helps to reduce water loss and resist microbial invasion. The shell proper is perforated with small openings, the pores, which run from the mammillary cores

Fig. 2.5. Reproductive system of the female fowl: mature ovary showing five rapidly developing oocytes. The largest oocyte, to the right of the picture, will be next to be ovulated and the rest will be released at about 24 h intervals. A collapsed, post-ovulatory follicle is visible at lower left.

right up to the cuticle. They are about 0.5 μm across and allow oxygen to diffuse into the shell and carbon dioxide and water to diffuse out. Research (Hughes *et al.*, 1986; Solomon, 1991) shows that egg shell formation can be affected by stress, thus suggesting that shell structure may be a non-invasive index of welfare.

Once oviposition begins, the sphincter between the shell gland and vagina relaxes, the shell gland contracts, the hen increases abdominal pressure and the egg is laid by passing through the vagina, cloaca and vent. After the membranes of the follicle have ruptured and released the ovum, its remnants form the post-ovulatory follicle. This has an important role – it secretes oestrogen and progesterone, which control the onset of pre-laying and nesting behaviour 24 h later, just prior to the laying of the egg (Wood-Gush and Gilbert, 1964, 1973).

Selection for egg number, together with *ad libitum* access to food, has transformed *G. gallus* from the jungle fowl that, under natural conditions, lays a clutch of 10–20 eggs, through primitive varieties such as Indian village fowls typically laying 40–50 eggs in a year, to the modern laying hybrid. Its very highly developed oviduct, together with its liver where lipid for the yolk is synthesized, produces over 300 eggs in 365 days. Initially ovulation occurs every 24–25 h but, as oviducal senescence takes place, the interval lengthens and sequences of eggs, which are separated by a non-laying day, become shorter.

2.11 Male Reproductive System

The main features are similar in all poultry species. The paired testes are in the dorso-central part of the body cavity, close to the kidneys and, in the average male

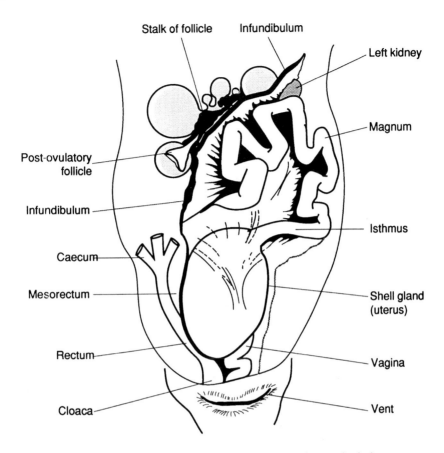

Fig. 2.6. Reproductive system of the female fowl: oviduct, down which the developing egg will pass from the fimbria of the infundibulum to the shell gland and then through the cloaca and vent (after King and McLelland, 1975).

domestic fowl (cockerel or rooster), at sexual maturity increase tenfold in weight from around 3 g to 30 g (Etches, 1996). They have a dual function: spermatogenic and endocrine. The sperm are produced over a 14-day period in the seminiferous tubules from diploid spermatogonia, with successive reduction (meiotic) divisions producing haploid spermatocytes, spermatids and finally spermatozoa. This differentiation is governed by hormone-rich secretions from the Sertoli cells. Suspended in the seminal fluid, the sperm are swept into a collection area called the rete testis, lined by ciliated cells which move them on into the epididymis. There the sperm are stored, mature and acquire motility and fertilizing capacity. The seminal fluid is a substrate providing energy and buffering capacity. The vas deferens transports the semen by peristalsis from the epididymis to the cloaca and also acts as a storage reservoir. When presented with a receptive female, the male mounts her and ejaculates through engorged phallic folds which protrude from the cloaca; in waterfowl, these folds are much more prominent and can be several centimetres long. Quail have a proctodeal gland and its secretions add a foamy consistency – it is speculated that this may help sperm metabolism by oxygenating the semen.

Cockerels previously kept in isolation can mate four times in 15 min, though around 50% of the semen is delivered in the first ejaculation (Etches, 1996).

The testes also have an endocrine function. The interstitial cells produce several androgens, the major hormone in blood being testosterone. At sexual maturity, luteinizing hormone (LH), a gonadotrophin produced by the anterior pituitary, is released and stimulates the output of testosterone. Plasma concentrations are maintained by a negative feedback loop – high concentrations of testosterone inhibit the output from the hypothalamus of gonadotrophin-releasing hormone (GnRH) which controls LH. As LH falls, so does the production of testosterone, and this in turn means that GnRH is again produced, to release more LH and thus raise androgen output once more (Etches, 1996).

2.12 Digestive System

The various species of domestic poultry, like all birds other than fossil species such as Archaeopteryx, lack teeth, having instead a horny beak with cutting edges. The tongue is heavily keratinized and is used for moving boluses of food within the oropharynx. Salivary glands are well developed; copious secretion from their numerous openings in the roof and floor of the oropharynx acts as a lubricant in swallowing the dry food typically consumed by gallinaceous birds. At the lower end of the oesophagus, just before it enters the body cavity, is a dilated sac, the distensible crop, which is especially well developed in seed-eating birds such as the domestic fowl. Food stored in the crop softens and swells during storage, and is then moved on through the remainder of the oesophagus into the proventriculus lined with glandular cells, where true digestion begins (Fig. 2.7). The next stage takes place in the thick-walled muscular gizzard, which is lined with a hardened membrane. Under natural conditions, birds eat small pieces of grit that localize here, and the considerable pressures exerted by grinding movements remove hard seed coats and break down the food to small particles, which are digested by hydrochloric acid and pepsin in the gastric juice. Most modern poultry feeding programmes supply easily digested food, such as mash or pellets, and thus grit is not provided. Domestic fowls still have a tendency, however, to peck up and swallow whatever hard, indigestible fragments are available, which may explain why birds kept on litter sometimes pack their crops full with wood shavings. Peristalsis moves the food along the duodenum, where bile and pancreatic secretions are added, into the jejunum and ileum where the majority of absorption occurs. The large intestine consists of paired caeca and a short straight section, probably homologous to the mammalian rectum. Breakdown of food by symbiotic bacteria occurs in the caeca, while water is reabsorbed in the rectum and cloaca. The faeces are voided from the rectum through the cloaca and the vent.

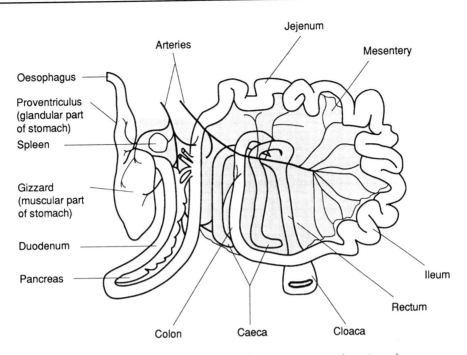

Fig. 2.7. Digestive system, showing the oesophagus, proventriculus, gizzard, duodenum, ileum, colon, caecum and rectum (after King and McLelland, 1975).

2.13 Cardiovascular System

Many species of bird have to perform strenuous and prolonged physical activities such as flying, running or swimming in harsh environments, and, accordingly, have evolved a high-performance cardiovascular system. Being homeotherms, with body temperatures in many cases higher than those of comparably sized mammals, an efficient circulation is also required to both conserve and remove heat. The avian system is similar in its basic design to that of mammals but is not a replica. Birds have larger hearts, bigger stroke volumes, lower heart rates and greater cardiac outputs than mammals of comparable mass (Grubb, 1983). The avian red blood cell differs from the mammalian cell in three major respects: it is nucleated, larger and ovoid rather than circular in shape. These latter properties may alter its flow characteristics through capillaries and affect viscosity (Smith *et al.*, 2000).

Selection for greater output and more efficient production in meat-type birds appears to have had deleterious effects on the effective functioning of the cardiovascular system. This has become apparent in the condition called ascites in broilers. First observed in fast-growing strains kept at high altitudes in cold conditions, it has since been reported throughout the world. Genetic selection for rapid growth and large muscle mass resulting in high metabolic requirements, coupled with a relative reduction in cardiovascular and lung function, has increased the probability of ascites (Decuypere *et al.*, 2000). It starts with hypoxaemia, increased blood viscosity and pulmonary vasoconstriction, which causes an increased load upon the heart. There is initial hypertrophy but eventually the heart fails to keep pace with the increased demand and ends in right-sided heart failure, extravasation of fluid into the peritoneal cavity and finally death. This occurs when the demand for oxygen from the tissues cannot be met by increased blood flow, and contributory factors include poor air quality, increased blood viscosity (in an attempt to carry more oxygen) and low temperatures resulting in vasoconstriction. It shows the importance of the breeder's role – and the need to select for adequate physiological support systems and not just focus on traits of economic importance.

2.14 Respiratory System

Most birds inhale air through a nasal cavity containing complex, scrolled turbinate bones lined with mucous membrane. This acts as a system for recovering water from the moisture-laden exhaled air, with some of its water vapour condensing on the cool nasal membranes and being reabsorbed. This mechanism limits evaporative loss and is obviously an important adaptation to hot, arid conditions. The larynx at the top of the trachea consists of four partly ossified cartilages supporting the glottis and prevents material such as food passing into the lower respiratory tract. In contrast to mammals, it plays no part in producing the voice. Because their forelimbs are specialized to form wings, birds have to use their bills for a wide range of functions, including investigative and aggressive pecking, preening and nest building. These functions require a long and mobile neck, which in turn necessitates a long trachea. To reduce resistance to air flow, it is considerably broader than in mammals of the same size, so has a correspondingly greater dead space. This is compensated for by much slower rates of breathing and much greater tidal volumes than in comparable mammals. At the lower end of the trachea, just before it divides to form the primary bronchi, is the syrinx, where the walls of the air passages are formed by membranes stretched between circular cartilaginous rings. The function of the syrinx is to generate sounds, produced by vibration of the membranes, the tension of which can be altered by muscles that attach to the cartilages above and below them.

The primary bronchi transport air to the lungs, which are in the dorsal part of the thoracic cavity, closely applied to the vertebral column and ribs, and are relatively small, with a volume only about a tenth of that of a similarly sized mammal (Fig. 2.8), inflating only slightly during inspiration. Unlike mammals, birds have no diaphragm separating the thoracic and abdominal cavities, while the lungs are composed entirely of tubes, rather than of tubes terminating in tiny sacs (alveoli) as in mammals. The primary bronchi divide into secondary bronchi and then into parabronchi, which in turn give rise to narrow passages called air capillaries. Gaseous exchange occurs mainly in the parabronchi and air capillaries. The latter have very small diameters, about 10 μm in swans down to 3 μm in small passerines. A network of blood capillaries runs along the walls of the air capillaries, separated from them by an extremely thin membrane, only 0.3 μm thick in the domestic fowl. There is a counter-current arrangement, with blood and air flowing past each other in different directions, which increases the effectiveness of oxygen transfer. The other major respect in which the respiratory system of birds differs from that of mammals is in their possession of air sacs. These arise directly from secondary

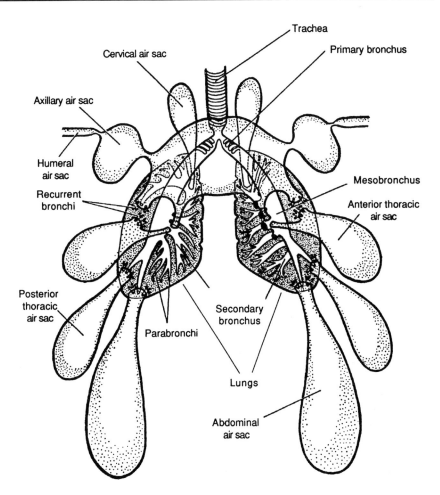

Fig. 2.8. Respiratory system, showing the relationship between the lungs, the parabronchi and the air sacs, which occupy both the peritoneal cavity and some of the bones (after Ede, 1964).

bronchi and extend forward into the cervical region, throughout the thoracic region and well into the abdomen. The walls of the sacs have a minimal blood supply and their function is clearly not that of gaseous exchange but, rather, to control air flow by inflating and deflating during inspiration and expiration. Diverticula from the air sacs extend into the cavities of a number of bones; in the fowl, these aerated bones include the sternum, scapula, humerus, femur, pelvis and numerous vertebrae and ribs. Such lightweight bones may be especially liable to breakage if birds are handled roughly.

The avian lung is much more efficient than the mammalian lung, partly because of more rapid gaseous diffusion, and partly because the area of exchange surface is relatively far greater in birds than in mammals. The domestic fowl, for example, has 18 cm^2 of exchange surface/g body weight, in contrast to about 2 cm^2 in the human. In addition, unlike the mammalian lung in which air flows in and out, the avian lung has a unidirectional air flow. On inspiration, air flows through the air capillaries into the anterior air sac and on expiration through the capillaries from the posterior sac (Schmidt-Nielsen, 1975). It is this high efficiency, resulting from gaseous exchange during both inspiration and expiration, that enables birds to maintain energetic activities such as flying for long periods with such small volumes of lung tissue.

2.15 Regulation of Water Balance and the Renal (Urinary) System

The regulation of body fluid composition differs markedly between birds and mammals. Birds do not have sweat glands, but have thin, well-vascularized skin, and most water is lost not through the kidneys but by evaporation through the skin and lungs. Evaporative water loss is normally about 50% of the total. This can be increased by panting and by feather raising to expose the skin to the air. When birds are dehydrated, water retention by the kidneys is so efficient that the proportion lost by evaporation can become 80% or even more.

The anatomy of the urinary system, too, differs in avians (Goldstein and Skadhauge, 2000). The glomeruli of the kidney are simpler but to compensate there are more of them, and there is no bladder: instead, the ureters empty directly into the cloaca. However, even though the avian kidney does not concentrate urine as effectively as in mammals, the arrangement whereby it is delivered into the hind gut allows retrograde flow into the colon and the caeca where secondary absorption of water and certain ions can occur. Another difference is that nitrogen excretion is mostly in the form of urates and ammonia, rather than urea as in mammals. The urates are not in solution but instead form a colloid and microspheres, allowing the urine to be extremely concentrated.

2.16 Immune System

As in mammals, the avian immune system is centred around the lymphocytes. These small round cells, with condensed chromatin and minimal cytoplasm, provide the killer cells, regulatory cells and memory cells of adaptive immune responses (Davison, 2003). They combat infections and prevent tumours. There are two main kinds of lymphocytes: the B cells (so called because they originate in the bursa of Fabricius) and the T cells, from the thymus gland. When a B cell is activated by exposure to an antigen, such as a pathogenic organism, it develops into a larger plasma cell, which in turn secretes antibodies, such as immunoglobulins, which help to neutralize the invader. T cells fall into two classes: cytotoxic types directly kill infected or tumour cells, while helper T cells direct and regulate other cells, such as B cells or macrophages, by releasing signal molecules. Lymphocytes also exert a memory function; they retain information about previous exposure to antigens, which allows the bird to mount a more rapid and effective response to a subsequent infection in the future.

Pathogens may be extracellular, such as bacteria, which are found circulating in the blood and other body fluids, and are thus easily discovered, attacked and destroyed. There are also intracellular pathogens, such as many viruses, which secrete themselves within the body's own cells or even in the host animal's genome, and are much more difficult to identify and destroy. This is where the major histocompatibility complex (MHC), which is a genetic region encoding complex molecules, has an important role. The function of the mammalian MHC is to collect peptides originating from intracellular pathogenic or neoplastic activity and to present them to receptors on the surface of the T cells. This in turn allows the T cells to identify infected or tumorous cells and to destroy them. Even though the MHC region is much smaller in birds than in mammals and the number of molecules produced is more limited, it still appears to have important functions, especially in disease resistance (Davison, 2003).

2.17 Integument

One of the clearest distinguishing features of birds is their feathers, which serve a number of functions: insulation, protection, waterproofing, cryptic coloration, sexual attraction and, not least, provision of the ability to fly. There are six types of feathers: the most obvious, which cover the outer surface of the body and include the flight feathers, are the contour feathers (Fig. 2.9). They grow from feather follicles, initially as a richly vascularized dermal core. Around this, an epidermal sheath develops to form the feather structure. This consists of a short basal tube, the calamus, which merges into the main shaft, or rachis. From this, the barbs protrude at an angle and in turn the barbules arise from them, engaging by means of hooklets called barbicels with barbules from the adjacent barbs. Contour feathers cannot function effectively unless they are in first-class condition – regular preening ensures that barbules remain firmly interlocked, distributes sebaceous secretions from the preen (uropygial) gland and helps to rearrange disturbed plumage. The other types of feather include down feathers, in which the barbules are not hooked, and which provide particularly effective insulation. Semiplumes have a similar structure but are much larger. At the base of each contour feather is a filoplume, which is richly innervated and probably provides proprioceptive input regarding the position of the feather. Bristles lack barbs, are found around the eyes and base of the beak, and have a tactile function. Powder feathers produce a white powder that helps to waterproof the contour feathers. Feathers deteriorate over time and are replaced by moulting. In domestic fowls, there are three moults in the first 6 months, with an annual moult thereafter during the autumn if they are kept under natural lighting conditions. The moulted feather is extruded by growth of the epidermal layer between the dermal core and the calamus, until it emerges cleanly. If, however, it is pulled out at any other time, for example by another bird, the procedure is painful, the dermal papilla is damaged and the follicle fills with blood. Beneath their feathers, birds have a thinner and more delicate skin than mammals, which can be easily damaged if the protection provided by the feathers is lost. In many species, including the domestic fowl, the lower part of the hind limbs is covered by scales rather than by feathers. On their heads, fowls, as well as some other species, have comb and wattles, composed of richly vascularized tissue, which probably both serve an ornamental function, signalling a bird's status and condition, and help to dissipate heat. Birds have no sweat glands, so their body temperature is regulated by evaporative cooling from the respiratory tract and by heat loss from the unfeathered areas. These two mechanisms are under behavioural control: panting increases evaporative cooling and wing raising exposes the poorly feathered sides of the body and under-wing area.

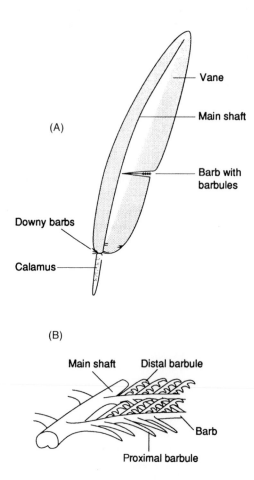

Fig. 2.9. Feather structure. (A) Primary flight feather from the wing. (B) Diagram showing how the barbs interlock by means of the hooked barbules (after King and McLelland, 1975).

2.18 Application of Biology to Housing Design and Breeding Strategies

This brief survey of avian biology has emphasized that we are dealing here with complex, highly evolved animals. The different species of poultry have been studied in detail and a great deal is now known about their structure and their environmental, physiological and behavioural requirements. This presents us with the opportunity to put this knowledge to good use. It is important to take into account all the features of poultry when designing production systems – some can be modified and controlled, but others may need careful consideration if problems are to be avoided – and also in deciding which traits to use as selection criteria when breeding.

3 Causes and Effects

3.1 Summary

- To understand behaviour, it is necessary to consider both its causes and its effects, by investigating its function, evolution, development and control.
- The function of behaviour refers to how it affects survival and reproduction. Effects may be environment-specific and, where behaviour is functional, this suggests that the bird is well adapted to the environment. Where it is not, it may cause problems for the animal itself, for other individuals or for the owner.
- Behaviour has evolved through natural selection, because certain behaviours increase the fitness of animals showing them, and is under strong genetic control.
- Domestication is a special, accelerated form of evolution. In the case of poultry, there are few behavioural differences between domestic fowls and their ancestors, and most of these are due to inadvertent rather than deliberate selection.
- The specific behaviour patterns of any individual reflect the interaction of its 'innate' behaviour with the environment in which it is reared.
- Damaging pecking, which appears to have a major genetic component modified by factors such as light intensity, group size, stocking density and environmental complexity, can be ameliorated by selection against the behaviour.
- Other behaviours have been altered by selection for desirable economic traits, such as the reduced mobility and concomitant placidity of broilers resulting from selection for improved food conversion efficiency.
- Learning is an important element in development of behaviour; it includes habituation, operant conditioning, associative learning and imprinting. Some of these can be put to use at a practical level.
- Control of behaviour includes physiological factors, e.g. oestrogen and progesterone help to trigger pre-laying behaviour, and mental processes, which involve elements of perception and possibly pleasure and suffering.

© M.C. Appleby, J.A. Mench and B.O. Hughes 2004. *Poultry Behaviour and Welfare*
(M.C. Appleby *et al.*)

3.2 Questions About Behaviour

Why do animals behave as they do? One of the most important contributions of behavioural science has been to point out that this 'question' actually includes several different sorts of question. We can illustrate this with an example.

If we ask a group of people 'Why does a cockerel crow?', we will get a variety of answers (Table 3.1). These answers may all be right, and most can be placed in two distinct categories. Some are concerned with the causes of behaviour, the factors that lead up it. These include the age of the individual, the concentrations of his hormones and the stimuli that he has received such as the increasing light of dawn. We might also mention the genes that affect the structure and function of his body, and the ways in which the internal and external stimuli (such as hormones and daylight) promote the muscular action necessary for the behaviour.

The other main category of answer relates to effects or potential effects of the behaviour. Crowing is part of a cockerel's competitive and territorial behaviour, and probably affects the breeding condition of females (as male vocalizations do in quail: Guyomarc'h *et al.*, 1981). In the longer term, therefore, it is likely to affect his breeding success.

It is not possible to separate causes and effects completely. Some answers to the question 'Why does a cockerel crow?' cannot be categorized in this way, e.g. the answer that cockerels enjoy crowing. If they do indeed enjoy crowing (such issues will be discussed in Chapter 8), then this may be both a cause and an effect of this behaviour (see Section 3.6). Nevertheless, clarification of the different sorts of questions that may be asked about behaviour is important.

Niko Tinbergen (1963), one of the founders of the scientific study of animal behaviour, categorized questions about behaviour in a different way, and his categories form the basis of this chapter. Tinbergen's categories each contain ideas relevant to both the causes and effects of behaviour. Concerning a particular behaviour, his questions are as follows.

- What is it for? (Function)
- How did it evolve? (Evolution)
- How did it arise during the life of that individual? (Development)
- What are the factors eliciting and controlling it? (Control)

Table 3.1. Possible answers to the question 'Why does a cockerel crow?'

It is dawn
It warns other males to keep away
He is a mature male with high testosterone levels
It encourages hens to mate with him
He enjoys crowing

3.3 Function

The question 'What is a particular behaviour pattern for?' could be understood in a number of ways, but in a biological context it is given an interpretation that is specific yet broad ranging: 'What are its likely effects on lifetime reproductive success?' The importance of this question can readily be understood in relation to the evolution of behaviour through natural selection, discussed in the next section. Its breadth becomes apparent when it is realized that in order to reproduce, an animal has first to survive and then to achieve all the precursors to reproduction such as attaining sufficient body condition. The function of a behaviour pattern, then, is defined as its likely effects on survival and reproduction.

Those effects will, of course, depend on the animal's environment. In the case of commercial poultry, some behaviours continue to be functional, i.e. to have positive effects on potential survival and reproduction, just as they did in the birds' ancestral environments. An example is feeding. Some behaviours that were once functional may now be less so: being aggressive to others who approach too closely was important to safeguard access to food in the wild but is problematic in a laying cage. Indeed, the function of egg laying is itself altered in commercial laying birds that have not mated, as the eggs are infertile. Lastly, some behaviours occur that would not have been seen in the wild. Some of these may be relatively easy to account for in functional terms, such as egg eating, which while surprising is certainly nutritive, but other examples such as stereotypic pacing are less so.

Understanding function, then, can be assisted by comparing the current environment with that in which the birds evolved – which has been called the environment of evolutionary adaptation (Mace, 1995; Barnard and Hurst, 1996). There are also three other approaches used to ensure that interpretation of function is not just guesswork. The first of these is detailed observation. For example, observation of cockerels in groups shows that high-ranking individuals crow more frequently than those of low rank (Fig. 3.1) and that when low rankers do crow, they are often attacked by high rankers (Leonard and Horn, 1995). This supports the idea that a function of crowing is advertisement of status as part of the competition between males. A second approach is comparison between species. Thus ganders do not crow: female and male geese vocalize with similar frequency, and mostly when they are gathering in groups. This suggests that the function of vocal advertisement in cockerels is associated with the fact that they are polygamous and must compete for mates, whereas geese form pair bonds (Chapter 6). Thirdly, functional arguments can be tested by experiment. Careful experiments were done on another form of vocalization by cockerels: they sometimes call when they discover food. The experiments showed that they are more likely to do so if there is a hen present than if they are either alone or accompanied by another male. The behaviour is termed 'food calling', but one function thus seems to be to attract females for possible mating (Marler et al., 1986).

Where behaviour is functional, this suggests that there is an appropriate matching between the behaviour and the environment, i.e. that the animal is adapted to the environment. This is to be expected in the environment of evolutionary adaptation, but one of the complicating factors in understanding function is that animals may adapt to a range of different environments: indeed,

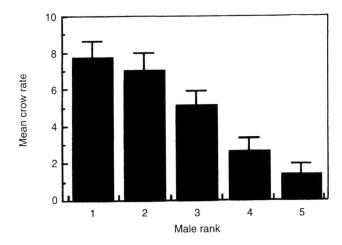

Fig. 3.1. Crowing by cockerels in groups of five: mean (and standard error) number of crows in a period of 50 min, from 24 groups (from Leonard and Horn, 1995, with permission from Elsevier).

adaptability is one of the characteristics that favoured domestication of certain species rather than others. Turkeys survive and reproduce in commercial conditions that are very different from those in which they evolved. So when behaviour is different in current conditions from that which would occur in the wild, this is not itself evidence of malfunction, of 'things going wrong'. Nevertheless, there is a broad category of behaviour, sometimes called 'abnormal behaviour' or 'behavioural problems', the main characteristic of which is that it would not be seen in the wild, at least at the frequency now observed. It is instructive to categorize such behaviour according to whether any harm results and if so to whom (Table 3.2). Where an individual's behaviour is a problem only for the owner or for other individuals, it may well still be functional. Egg eating, for example, is always initiated by accidental breakage of eggs, but once birds have experience of eating broken eggs they may learn to break more themselves: a very functional way of obtaining food. However, behaviour that causes problems for the individual concerned, and some other behaviour with no apparent function such as stereotypic pacing (Chapter 9), does indicate a failure to adapt to current conditions – a limit to adaptability.

Behavioural problems are sometimes referred to as 'vices', but this term is inappropriate as it suggests that the causes are intrinsic to the birds, that the birds are 'to blame' rather than the environment. Further, it suggests that the problems are inherent and therefore insoluble. On the contrary, appropriate management techniques can often reduce the effects of such behaviour when it occurs. Even more importantly, good management can help to prevent it occurring.

One mistake that is sometimes made in considering the functions of behaviours is to assume, usually implicitly, that an animal understands the long-term effects of its own behaviour: that a goose mates because she wants goslings or that a pheasant avoids a fox because he understands and fears death. The instincts for sex and avoidance of novel animals such as foxes do not require understanding of long-term

Table 3.2. Behavioural problems.

Problem for owner
  Egg eating

Problem for other individuals
  Aggressiveness
  Feather pecking
  Cannibalism

Problem for animal performing the behaviour
  Hysteria
  Excessive gregariousness, leading to suffocation

Usually not directly harmful
  Stereotypic pacing
  Vacuum nesting
  Vacuum dust bathing

effects, and the lack of such understanding is demonstrated by the sort of failure to adapt just discussed. For example, the fact that tame hens crouch in sexual lordosis when a human approaches suggests that they do not understand why they are doing so. This does not mean that animals have no understanding at all: control of behaviour may involve animals understanding at least its short-term effects (see Section 3.6 and Chapter 8).

Another misunderstanding of functional arguments is seen in suggestions that an animal behaves in a certain way 'for the good of the species' or 'to perpetuate the species'. That is not how natural selection acts. Natural selection acts on genes and on the individuals that carry them, not on species as a whole. Genes that affect behaviour in such a way that individuals carrying them are more successful at reproducing than individuals carrying alternative versions will increase in the local population, compared with those alternative versions. Nothing in that process involves the wider population that we as observers call a species; indeed, it may cause change in the local population and, in due course, the splitting-off of a new species.

A final point to make about the function of behaviour is that the action of natural selection on gene frequencies does not just involve an individual and its own reproduction but also other members of the local population, particularly its relatives. This is how natural selection has produced social behaviour. First, individuals often help each other to mutual benefit that may be either simultaneous (as when pheasants roost in contact for shared bodily warmth) or turn-and-turn-about (as when a quail helps others by alarm calling, and is itself helped similarly on other occasions). Secondly, social behaviour is especially common between relatives, because relatives have genes in common by inheritance from common ancestors. Thus the genes influencing a turkey hen's care of her poults are likely to be perpetuated because some of the poults will inherit them from her. Social behaviour and its functions are discussed in Chapters 5 and 6.

Utilization, manipulation and prevention of specific behaviour patterns of poultry usually involve understanding how those behaviours are controlled. However, knowledge about function is important to put this understanding into a broader context. It is also important as a basis for understanding evolution and artificial genetic selection, to which we turn next.

3.4 Evolution and Domestication

Behaviour, like the other characteristics of organisms, has evolved through natural selection, because it is strongly affected by genes. Under natural conditions, those individuals whose behaviour best equips them survive and leave most offspring, which tend to inherit their parents' behaviour. The process also affects others with whom they interact, particularly relatives, as discussed above. For example, jungle fowl evolved in the rainforest of South-east Asia, and their behaviour patterns are those typical of ground-dwelling birds. They live in groups, spend much of their time in cover and show appropriate species-specific protective responses to predators, remaining silent, crouching and freezing to overhead predators, while calling and running or flying away from ground predators.

Reproductive success is also affected by the environment, and different populations are subject to different environmental pressures. So natural selection favours different behaviour in different populations, with the result that populations differ in behaviour as well as in other characteristics. If these differences become great enough, then the populations may become established as separate species.

The process of domestication can be regarded as a special, often accelerated, form of evolution. Together with the predictable effects on production traits such as more rapid growth rate, greater body size and increased egg output, domestication has resulted in characteristic changes in behaviour (Schutz and Jensen, 2001; Schutz *et al.*, 2001; Price, 2002). However, few of the differences in behaviour between domesticated poultry and their ancestral forms are due to deliberate selection. Some patterns that first evolved in wild birds, such as nesting behaviour and anti-predator responses, remain almost unchanged in modern poultry, presumably because these behaviours were widespread and stable in the genotype and there has been no selection against them. Other behaviour patterns have been strongly selected against and have almost disappeared from modern hybrids. An example is broodiness in hens, which was inadvertently selected against as a consequence of selecting egg-laying hybrids on the basis of high egg number. Individuals that became broody during the test period laid fewer eggs and were withdrawn from the breeding population. Breeds such as bantams, which have not undergone intensive selection for egg output but instead have been kept for their appearance or other characteristics, have retained broody behaviour.

It is important to note that no behaviour pattern has been eliminated completely in the process of domestication, not even broodiness, nor has any new behaviour pattern been produced. The behavioural repertoires of poultry species are therefore identical to those of their wild ancestors. This is illustrated by the fact that poultry can become feral, surviving and breeding successfully in the wild

(Duncan *et al.*, 1978; Savory *et al.*, 1978). However, artificial selection has resulted in other differences such as in frequencies of different types of behaviour.

In an elegant series of experiments on ducks, wild-type mallards and domestic Aylesbury ducks were hatched and reared under identical conditions so that any variation between the ancestral and the domestic types could be attributed to genetic rather than environmental causes (Desforges and Wood-Gush, 1975a,b). There was less aggression and more tolerance towards other individuals in the domestic ducks, their flight distance was much less when humans approached and they habituated more quickly to novel stimuli (Fig. 3.2). They were also more willing to eat unfamiliar foods. Changes such as these, which are also seen in many other domesticated species, are unlikely to be due to conscious selection by humans, even though they are advantageous as far as management and feeding are concerned. Instead, they probably occurred because individuals showing these traits were best suited to the husbandry, dietary and housing conditions of domestication and left more offspring than those less well adapted.

In a few cases, there has been selection for a particular behaviour. In the early days of domestication, it is likely that fowl were kept mainly for cockfighting, so cockerels were carefully selected for maximum aggression, with only those that won contests being retained for breeding. Modern cockerels crow much more frequently than jungle fowl. This is believed to be a relic of their involvement in religious ceremonies such as those of Zoroastrianism, in which crowing cocks played an

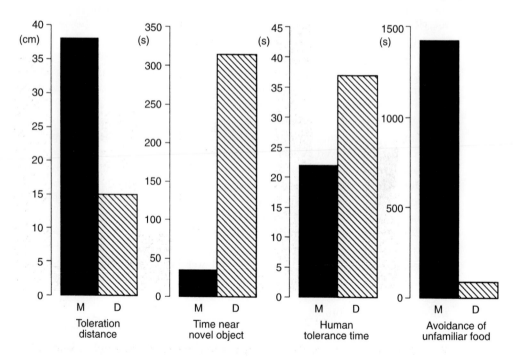

Fig. 3.2. Compared with wild-type mallard (M), domestic ducks (D) tolerated other individuals nearer to them without aggression, spent more time close to a novel object, took longer to move away from a human being when released and fed sooner when food was presented in an unfamiliar container (from Desforges and Wood-Gush, 1975a,b, with permission from Elsevier).

important role and were, presumably, selected on the basis of loud, long and frequent crowing (Wood-Gush, 1959a).

Most selection for behavioural traits has been on an experimental basis, and examples are given in the appropriate chapters. It should be said, though, that there is increasing interest in such experiments. The importance of understanding the genetic components of behaviour is that certain traits that are today regarded as undesirable might be reduced or eliminated by selection. Possible examples would be fearful or panic responses, cannibalism, feather pecking and inter-individual aggression in laying flocks (Mills *et al.*, 1997a; Jones and Hocking, 2000). An example will show both the potential and the problems of such selection. There are substantial differences between strains of laying hens in the incidence of pre-laying agitation and pacing in cages (Mills *et al.*, 1985a), as Fig. 3.3 shows. Pre-laying agitation and pacing have been shown to indicate frustration (Duncan, 1982). The various elements of this behaviour are under partial genetic control and so they could, at least in principle, be selected against (Braastad and Katle, 1989; Heil *et al.*, 1990). Whether this would be beneficial for welfare is, however, an open question. The selection might be only against the outward display of frustration, in which case the underlying distress might continue.

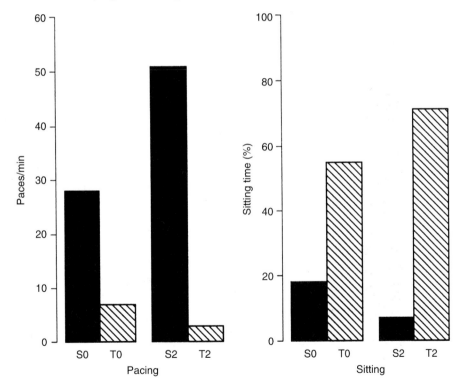

Fig. 3.3. Before genetic selection, during the 10 min period before laying, S-line (S0) hens paced more than T-line (T0) hens and spent less time sitting. After two generations of bidirectional selection, for pacing in the S-line (S2) and for sitting in the T-line (T2), these differences were exaggerated (from Mills *et al.*, 1985a, with permission from Taylor & Francis Ltd).

There is also interest in controlling for other aspects of welfare during selection. As one example, selection for high growth rate in broilers has led to an increase in leg disorders (see Section 8.6). Some of these disorders have a strong heritable component and can be reduced without much effect on growth (Sorensen, 1989).

The fact that selection for certain traits often produces changes in others has had effects on behaviour. The most obvious of these is the increased placidity in strains selected for meat production, especially broilers. The same effect accounts for differences in flightiness between medium weight hens such as the Rhode Island Red, originally selected as a dual-purpose breed for both meat and egg production, and the more nervous lightweight breeds such as White Leghorns, selected only for eggs. It is reasonable that higher growth rate should be associated with a calmer disposition, but the actual mechanism is not known.

3.5 Development

Although genetic factors play a major role in behaviour, the specific behaviour patterns shown by an individual animal result from the interaction of those factors with the environment experienced by that animal during its lifetime. So while birds possess a relatively large proportion of species-specific behaviour patterns, 'innate' or 'pre-programmed', these are modified in their expression by interaction with the environment. Maternal imprinting provides a good example of this interaction. Chicks and ducklings have an innate tendency to imprint on the first moving object they see and to treat it as a parent (see Section 5.4). Thus, the primary tendency to follow is inherited, but they learn what to follow. Furthermore, there are inherited constraints on what they may learn. Chicks are more likely to imprint if the object is about the size of a hen than if it is much smaller, if the object emits calling sounds like those of a hen than if it is silent, and if it is patterned or moving than if it is plain or stationary. There are interesting between-species differences in ducks, which are clearly adaptive: mallard, which nest on the ground, imprint to visual stimuli, whereas wood duck, which nest in holes in trees and rear their young in the dark, imprint to auditory stimuli (Klopfer, 1959).

Other behaviour also shows this pattern of an innate base modified appropriately by environmental factors. Soon after they have hatched, chicks show an innate tendency to peck at a wide range of stimuli around them. At this stage, they will peck equally at grains of sand or at particles of food (Hogan, 1971). However, as time passes, the pecking at sand wanes while the pecking at food strengthens in response to the positive nutritive feedback. Another example is nesting behaviour, which appears as a fully organized collection of behaviour patterns on the very first occasion when a bird lays an egg. However, the precise nature of its expression thereafter is greatly modified by the surroundings in which the bird finds herself. Birds with access to adequate nest sites display a full repertoire. For hens in cages, the investigative phase may be replaced by a prolonged period of stereotyped pacing or escape behaviour, while the final sitting phase may be almost entirely absent (see Section 6.11). A final example is provided by the fearfulness which many birds display towards humans. The level of this fear is very high in jungle fowl, but selection during the course of domestication has considerably reduced it in modern strains. It can be reduced still further by environmental factors (Fig. 3.4), such as regular human handling (Jones, 1987b) or exposure of young chicks to enriched environments (Jones, 1989).

One of the major processes in development of behaviour, then, is learning. Learning can be defined broadly as 'internal change causing adaptive modifications in behaviour as a result of experience' (Thorpe, 1951). The simplest form of learning is habituation, in which a response, after a number of repetitions, gradually wanes. An example of this is chicks' responses to an object passing over them. At first, they show an anti-predator response: they stop what they are doing, crouch and freeze. If the action is repeated a number of times without any unpleasant consequences for the chicks, the duration of freezing becomes shorter and eventually they ignore the stimulus altogether. This is an innate response modified by the birds' experience. However, the response will be restored by the passage of time or if there is any appreciable change in the nature of the stimulus. Habituation is a useful reaction. It allows animals to adapt quickly to prominent, harmless features of the environment that initially caused them disturbance. This effect is probably even more important in commercial conditions than in the natural situation, e.g. in reducing the response to humans, who are initially seen as frightening.

Classical conditioning is the type of learning first described by Pavlov, in which animals learn to associate different stimuli that appear close in time. Dogs normally

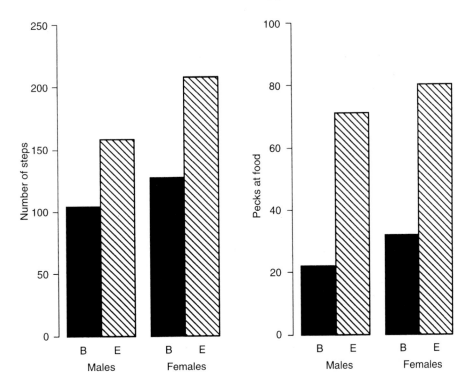

Fig. 3.4. Male and female domestic chicks reared in a barren (B) rather than an enriched (E) environment move around less and peck less at food when they are placed in a novel enclosure for 15 min (Jones, 1982).

salivate in anticipation if presented with food such as meat. What Pavlov found was that dogs would learn to salivate in response to a stimulus such as the ticking of a metronome, provided the metronome was activated on a number of occasions immediately before a piece of meat was provided. In the same way, hens learn to respond to a previously neutral stimulus if it occurs sufficiently often in association with another stimulus that itself causes a predictable response. For example, in one experiment, hens were given a fright by suddenly inflating a balloon close to them; they jumped up, flapped their wings and squawked. If a lamp was switched on a few seconds before the balloon was inflated, the birds soon became disturbed by the lamp on its own and often showed the escape response even if the balloon was not inflated (Rutter and Duncan, 1991). This kind of associative learning is obviously adaptive under natural circumstances: animals that learn to avoid stimuli associated with the presence of predators are less likely to be caught and preyed upon.

Another form of learning is operant conditioning, in which animals learn to carry out behaviour to obtain food, water or some other desirable consequence. Behaviour followed by a pleasant outcome is more likely to occur again, whereas that with neutral or unpleasant consequences is less likely to be performed in future. The consequence of the operant behaviour is described as 'reinforcement' and can be positive or negative. Thus, hens can be trained to peck at a small disc to operate a feeding device that gives them access to food, and this is, of course, positive reinforcement. If, however, pecking the disc instead has unpleasant consequences, such as exposure to a stimulus they find frightening, they quickly learn to stop pecking. This effect can be described as punishment. This type of learning again is adaptive. Poults that at first will peck at a wide range of small objects increase their pecking at food particles because eating them has pleasant consequences, whereas they soon cease pecking at sand because its consumption does not provide them with positive reinforcement.

Imprinting, already mentioned above, is a special form of learning during which a newly hatched chick or duckling may learn the characteristics of a parent. It occurs during a sensitive period that lasts 24–36 h after hatching. The learning may be very specific: domestic chicks exposed to a human being for 15 min can distinguish between that person and a stranger (Gray and Howard, 1957). Under natural conditions, imprinting is adaptive because the hatchling bonds to its mother and follows her closely, reducing the risk of separation during a vulnerable stage of its life. Poultry under commercial conditions are taken from the incubator trays shortly after hatching, swiftly handled and sorted, placed in boxes and transported to a rearing unit, where they are brooded in large groups. They do not see their mother and cannot imprint on her, while any visual contact with humans is very brief. If they imprint at all, it is on other chicks in their group, the only moving objects they see for any length of time during the sensitive period. This imprinting may help to explain why isolating young chicks from their group mates causes considerable distress, generally manifested as prolonged peeping calls.

Our knowledge of poultry behaviour and the factors influencing it can serve a practical purpose, as the early experience of birds may have a considerable effect on their subsequent behaviour. For example, enriching the environment of both domestic chicks and quail chicks by exposing them to a range of novel objects or stimuli makes them more able to resist disturbance and stress later in life. In particular, they seem to be less fearful (Jones, 1987b; Jones *et al.*, 1991). This was put to practical effect by one broiler producer, who regularly walked through his flock of young chickens banging a metal can with an iron bar. This habituated them to disturbance and made them easier to catch when the time came to empty the house (I.J.H. Duncan, personal communication).

In another example, pullets given access to perches during the rearing period lay fewer eggs on the floor when they are subsequently housed in multi-level systems such as percheries or aviaries (see Section 6.10). This is because perch-reared birds become accustomed to moving around in three dimensions and thus are able to find and reach raised nest boxes without difficulty.

Environmental factors are frequently important in influencing undesirable behaviour. The control of pecking and cannibalism still poses major problems, but a number of influential factors, such as complexity of the environment, light intensity, group size, stocking density and disturbance, have been identified (see Section 5.10; Hughes and Duncan, 1972) as Fig. 3.5 shows. Thus the conditions under which pecking is likely to occur can be described, even if the precise factors that trigger a particular outbreak cannot be identified. Because there are large differences between strains, genetic selection is also likely to be effective in modifying this behaviour.

3.6 Control

Interaction of genes, body structure (including neurophysiology) and environment is also, of course, critical in short-term control of behaviour. We will first consider physiological control; environmental controls will be discussed in subsequent chapters.

One of the important functions of behaviour is to help maintain the constancy of an animal's internal environment; a good example of this is ingestive behaviour. The control of feeding behaviour is increasingly understood and it is clear that parts of the hypothalamus, an area at the base of the brain, are concerned with the initiation and termination of feeding. Activity in the lateral hypothalamus increases if animals are kept without food for long periods, in response to physiological changes such as an empty gut or a reduction in circulating metabolites such as glucose. This part of the brain can thus be regarded as a 'feeding centre' and if it is artificially stimulated the animal begins to eat. The lower-central part of the hypothalamus, on the other hand, is a 'satiety centre' and animals stop feeding when it is active. It responds to stimuli such as a full gut.

A number of hormones have important roles in the control of behaviour, especially that connected with reproduction, aggression and responses to stress. This control is complex, generally involving several stages. For example, substances called releasing factors are secreted by the hypothalamus and pass to a gland below the brain called the anterior pituitary. There they release hormones into the blood. An example of one of these hormones is adrenocorticotrophic hormone (ACTH),

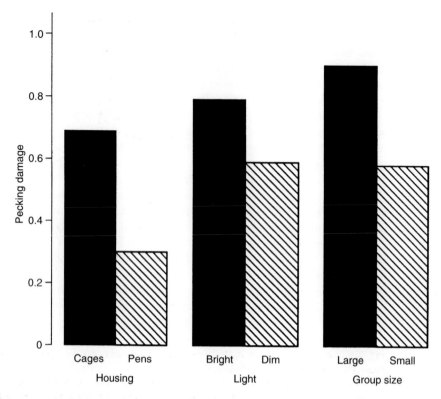

Fig. 3.5. Pecking damage (on a scale from 0 to 4) was greater in cages than in pens, when light intensity was bright rather than dim and when birds were caged in large rather than small groups (Hughes and Duncan, 1972).

which stimulates the adrenal glands to produce another hormone, the steroid called corticosterone (Fig. 3.6). This acts on the brain to influence behaviour by changing perception.

Pre-laying behaviour is under hormonal control. Wood-Gush and Gilbert (1975) showed that oestrogen and progesterone released from the ovary around the time of ovulation initiate the sequence of pre-laying behaviour about 24 h later, when the egg is ready to be laid, which results in nesting behaviour that is terminated by the laying of the egg.

However much we know about the physiological control of behaviour, though, it would be misleading to suggest that we are anywhere near being able to predict from such knowledge precisely what animals will do. A 'mechanical' approach is still inadequate for understanding the control of complex behaviour, particularly because animals do not normally just have one behavioural option available but instead many simultaneously. Pre-laying behaviour, for example, may start and stop several times, due to many factors including whether food is available and how long it is since the bird last ate (Freire *et al.*, 1997), whether there is any threat of predation, and so on. It is usual to address this complexity by describing the mental control of behaviour. Often this is done casually, without fully identifying the implications. For example, in the previous section, we described behaviour as having

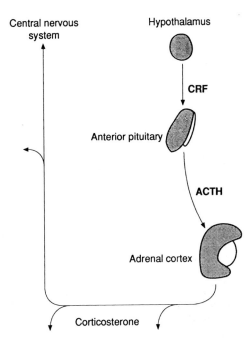

Central nervous
system

Hypothalamus

CRF

Anterior pituitary

ACTH

Adrenal cortex

Corticosterone

Fig. 3.6. Hormonal control of stress response. Corticotrophin-releasing factor (CRF) is secreted by the hypothalamus and causes the anterior pituitary to release adrenocorticotrophic hormone (ACTH) into the circulation. ACTH stimulates the adrenal cortex to produce corticosterone, which among other actions can influence the activity of the brain.

'pleasant' or 'unpleasant' consequences, leading to 'positive reinforcement' or 'punishment', respectively, and hence to the behaviour increasing or decreasing. There are twin dangers to a casual approach. First, we may make assumptions about what is pleasant or unpleasant for the animal solely on the basis of what is pleasant or unpleasant for us: this can be described as uncritical anthropomorphism. Secondly, we may make the circular and therefore unprovable assumption that the consequences of the behaviour must be pleasant, because the animal is seen to repeat the behaviour.

We can avoid both these dangers if our discussion of mental processes is both explicit and cautious. Thus the study of hens' pre-laying behaviour by Freire *et al.* (1997) used the concepts of pre-laying motivation and feeding motivation or hunger. Motivation can be described as the stage in the processing of information (e.g. in a stimulus–response sequence) that is perceived by the animal (Appleby *et al.*, 1992). Looking at pre-laying behaviour in this way helps to explain its occurrence, including interruptions and the effects of those interruptions. It would obviously be wrong to assume that what was described as hunger in the birds is exactly the same as human hunger; however, it would also be a mistake to avoid any mention of hunger in animals, as this would limit our understanding of behaviour. Furthermore, there is an almost universal consensus that animals are not machines, that they do have mental processes that are similar in at least some respects to our own, including feeling pleasure and suffering. Again, 'perception' by an animal is not the same as human perception, but must have at least some elements in common. Cautious, explicit discussion of mental processes in animals may be described as careful anthropomorphism: using aspects of our own mental processes as models for what is happening in animals.

The likely involvement of pleasure and suffering in the control of behaviour (Fraser and Duncan, 1998) leads to the overlap between causes and effects of behaviour mentioned earlier. When a newly hatched poult eats a food particle and this has pleasant effects, it learns to eat more such particles rather than, say, sand. It can therefore be said that the poult anticipates the pleasure of eating, and anticipation of the effect of behaviour may itself be a cause. Similarly, when a pheasant runs away from a fox, fear of the fox is a cause of the behaviour, but reduction of that fear can be considered as both a motivation for running and an effect of it.

It will readily be appreciated that consideration of mental processes in animals, such as pleasure and suffering, has implications for animal welfare. Indeed, this consideration is central to one major approach to welfare, that of animal feelings. This approach will be discussed in Chapter 8.

The remaining chapters of Part B cover key aspects of the behaviour of poultry and their interactions with the environment, opportunities for using them to best advantage and problems that may arise.

4 Maintenance

4.1 Summary

- Wild poultry are active, daily moving long distances within their home range to forage, while at night they roost. Except in waterfowl, flight is infrequent.
- Chicks peck at small objects and, over a few days, learn to peck nutritious items. Chicks initially fail to recognize water, but tend to peck at shiny surfaces and, once their beak is immersed, soon learn to drink.
- In floor systems, birds spend much time foraging: pecking and scratching. On wire floors, they instead manipulate the food in their troughs. Food wastage can be reduced by trough design or by wire grids, but absence of a manipulable substrate in cages can cause feather pecking, which may also occur in floor systems if the littered area is small.
- Food requirements are affected by wastage, energy needs, nutrient density and diet palatability. Chickens have the ability to choose a nutritionally adequate diet from a range of different foods, and they have specific appetites for key nutrients.
- Feeding occurs in bouts and shows a clear diurnal rhythm. It is controlled by both hunger and satiety mechanisms, with signals from the gut playing a key role.
- Birds tend to feed as a group, even in single cages. In large flocks, enough space should be provided to allow most birds to feed simultaneously, otherwise they could be frustrated and production may decrease.
- Drinking is closely associated with feeding. Poorly designed or sited drinkers, high mineral diets and high densities can cause water spillage, while overdrinking is linked to stress or food restriction. Both lead to wet litter and health problems.
- Many cages for laying hens provide space less than or equal to their body area. Movement is severely restricted, wing flaps (seen regularly on deep litter) are

absent and bone strength is reduced. Whether birds need to move large distances is less clear.

● Comfort movements such as preening, dust and water bathing, wing flapping and feather ruffling are important to keep the plumage in good condition. The incidence of these behaviours is influenced by availability of space and substrates. They decrease with crowding and are much less frequent in cages.

● Regular periods of light and darkness are important to allow the development of a diurnal cycle and to allow resting and sleeping. Normal resting and sleeping are also promoted by provision of a suitable roost area, and chickens are strongly motivated to reach and use perches.

4.2 Natural Behaviour

Maintenance behaviours are those behaviours through which animals sustain their physiological equilibrium. They include feeding, drinking, resting, comfort behaviours, such as those involved with care of the plumage, and the patterns of activity associated with these behaviours. Wild and feral poultry are active birds, sometimes moving many kilometres during a day (Mench and Keeling, 2001). They usually have a fairly well defined home range and become familiar with this area and know the best places for feeding and resting. How much of this range is used daily will depend upon factors such as the distribution of food, water, flockmates, predators and, for fowl, desirable roosting sites. Roosting at night occurs in regularly used bushes or trees, while shorter rests in the day are usually also off the ground but in more variable locations. Flight is infrequent in fowl except when going up to such resting places or descending and, although chicks are often very mobile, adults usually walk unless running or flying is necessary. Waterfowl are similarly active, either in water or on land, and generally fly more. As a result, they also often use larger areas from day to day than do ground-dwelling birds.

Much of the time spent active is devoted to foraging for and consuming food (e.g. Dawkins, 1989; Deeming and Bubier, 1999). Under natural conditions, the diet of most poultry species is a very mixed one. For example, jungle fowl and feral fowl consume seeds, fruits, herbage, invertebrates and even carrion (McBride *et al.*, 1969). The diet of other galliforms is similar. Dabbling ducks such as the mallard, the ancestor of the Pekin, consume a variety of aquatic plants and invertebrates, while geese and Muscovy ducks browse for forage and invertebrates in grass, although Muscovy will also feed in shallow water. Adult ostriches are almost exclusively vegetarian, consuming grasses and the leaves, flowers and fruits of plants, although they have also been observed to swallow dry bones (Deeming and Bubier, 1999). Diets change with age as nutritional requirements change: a wide survey of 21 galliform species showed that 16 of them consumed mostly animal food in the first 2 weeks of life, whereas, by 8 weeks of age, 20 out of 21 species were subsisting mainly on plant material (Savory, 1989).

Even when fed concentrated feeds, poultry will still consume herbage if it is available. Medium hybrid hens in small flocks on free range given *ad libitum* mash were found to consume a considerable quantity of grass from the pasture, about 50 g per day (Hughes and Dun, 1983). The tendency of ostriches to continue consuming grass even when given concentrated feed is a problem in commercial production systems because it leads to the range becoming denuded of cover (Deeming and Bubier, 1999). Poultry also eat indigestible matter such as sand, fine grit or stones. Some of this is retained in the gizzard and, in free-range birds and others that are fed large food items, helps to grind the food.

In the variable environmental conditions encountered in a natural habitat, it is important for birds to keep their plumage in good condition, and this is achieved by frequent preening and other comfort behaviours. In waterfowl, this behaviour also acts to waterproof the feathers. Wild birds nearly always look well groomed, and wild and feral poultry are no exception in this respect.

4.3 Development of Feeding and Drinking Behaviour

Newly hatched precocial chicks do not have an innate ability to recognize food, but they possess a strong propensity to peck at small particles, both nutritious and non-nutritious (Hogan, 1973). Much pecking is done with a closed beak and does not lead to the ingestion of the object, at least during the first few days after hatching when the chick is receiving nutrition from the remains of the yolk sac. As the chicks explore and learn to respond to the consequences of consuming different items, however, pecking at inedible particles, such as sand, declines and pecking at food increases.

Under natural conditions, chicks' attention is directed toward food by the pecking and vocalization of the hen (Collias, 1952; Evans, 1975). The only production system in which this would now occur is one in which chicks are naturally brooded, e.g. a small farmyard flock. However, even without the stimulus of a hen, chicks can be stimulated to feed by making a sharp tapping sound similar to the sound made by a pecking hen (Suboski, 1987), a method that is often used to attract chicks to feeders. Chicks feed as a group when possible, and the amount of food eaten increases when companions are present (Tolman and Wilson, 1965). That feeding is socially facilitated (Section 5.6) in this way means that chicks are able to learn to feed from one another when kept under commercial conditions. Some feeding-related problems can also be addressed by taking advantage of the social nature of this behaviour. For example, putting young turkey poults, which sometimes suffer from a condition known as 'starve out' during which they fail to start eating, with a broiler chick that is feeding well can be effective in encouraging the poults to eat (Savory, 1982). Placing conspicuous and attractive stimuli near the feeder, such as flashing coloured lights, can also stimulate feeding by poults (Lewis and Hurnik, 1979).

As is the case for food, young chicks are initially unable to recognize water. They have, however, a tendency to peck at shiny surfaces. This results in them pecking at a pool of water, and once their beaks are immersed they begin to learn to drink; the characteristic movement during which the head is raised and swallowing occurs is innate. Chicks under commercial conditions have some difficulty in learning to peck at nipple drinkers; this movement has to be learnt. For this reason, the pressure in the system is often increased for the first few days, so that water drips slowly from the drinkers, thus encouraging the chicks to peck at the shiny drops.

4.4 Foraging Behaviour

There is increasing evidence that foraging behaviour is important to poultry. Under semi-natural conditions, jungle fowl, even though they are fed regularly, allocate a large proportion of their time to foraging activities (Dawkins, 1989). Pecking and scratching are common components of foraging behaviour for chickens and other galliforms, and they commonly scratch with both feet then move quickly backwards, pecking at anything edible exposed by the scratching. This is considered to be the appetitive component of feeding behaviour, while the picking up and swallowing of the food is consummatory behaviour. Pecking and scratching are performed in loose litter, if it is available. In non-cage systems, such as covered strawyards or deep litter, foraging in the form of scratching and pecking occupies a considerable proportion of the day, from 7 to 25% (Fig. 4.1; Gibson *et al.*, 1988; Appleby *et al.*, 1989). If there is edible matter in the substrate, pecking and scratching are considerably increased, and it used to be common to scatter some grain in deep litter houses and strawyards for this reason, to help keep the litter in good condition.

In conventional cages, birds do not have access to loose material but instead spend a substantial proportion of time either feeding or manipulating the food in the trough with their beaks. The manipulation takes two main forms: food is either drawn towards the bird and piled up at the back of the trough or it is flicked back and forth with vigorous beak movements, some of it ending up outside the trough and being wasted (Fig. 4.2). These movements probably represent the appetitive components of foraging behaviour, which the birds carry out in the food because it is the only substrate to which they have access.

Food wastage through manipulation can be economically important, so a number of techniques have been adopted commercially to minimize it. In countries where it is allowed, beak trimming is sometimes used to prevent feed wastage, since when the beak tip is removed feed can no longer be caught under the beak hook and flicked out of the feeder. Also, birds beak-trimmed at later ages may experience pain in the beak stump, which decreases beak-related activities, including foraging behaviour. Feeder design is very important. Wastage can be minimized in several ways, e.g. by placing a wire grid at the level of the food so that the birds have to peck through the spaces in order to feed, or by having a spiral along the bottom of the trough that prevents flicking, or by using relatively deep, narrow troughs with a shallow depth of food replenished by an automatic chain or other conveyor running in the base of the troughs (Fig. 4.3). If the ability to perform appetitive behaviour is important, then it is possible that wastage-reducing methods could be a source of frustration to caged hens. To safeguard welfare, it may be necessary to provide them with a suitable substrate, as will now be required in the EU in furnished cages (Commission of the European Communities, 1999).

It is not clear how motivated birds are to obtain loose material in which to peck and scratch. However, motivation for such behaviour is probably not stimulus bound, since it is an integral part of feeding behaviour even where that seems to be inappropriate, e.g. chickens scratch even if feeding on freely available food contained in a metal trough. This tendency has actually been used to advantage in battery cages, in a particularly good example of how management can utilize behaviour. If a strip of abrasive tape is fixed to the manure deflector behind the

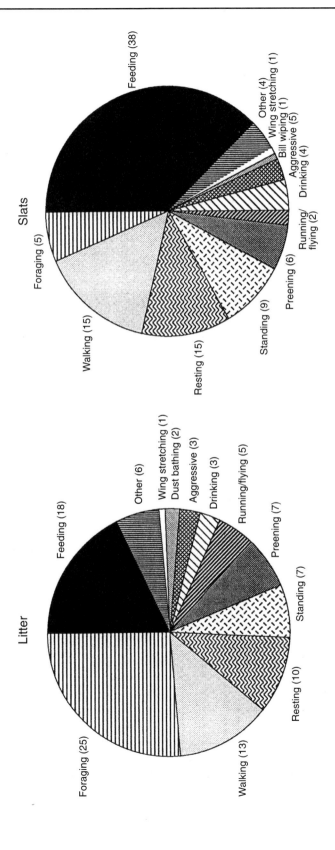

Fig. 4.1. Behaviour of laying hens in a deep litter house: proportion of time spent in different activities on the litter (left) and on a raised, slatted area (right) (from Appleby *et al.*, 1989, with permission from Taylor & Francis Ltd).

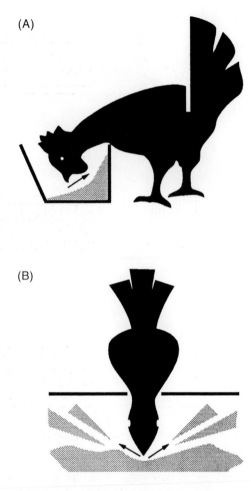

Fig. 4.2. Two types of food manipulation performed by caged hens, which can result in food being pulled or flicked out of the trough, resulting in wastage. (A) Beak drawn back towards body. (B) Beak flicked from side to side, scattering food.

food trough, scratching by the hens prevents overgrowth of their claws and the damage and injury that this can produce (Tauson, 1986). This adaptation of cages is now compulsory in Sweden, and the EU standards now require similar claw-shortening devices to be installed in furnished cages (Commission of the European Communities, 1999).

Whether or not birds are highly motivated to obtain loose material, the absence of varied or manipulable substrates in systems leads to other problems. Perhaps most importantly, it probably contributes to the development of feather pecking and cannibalism (Section 5.10) in cages or wire floor systems: in one experiment, pullets in pens that were deprived of such substrates showed an increased frequency of redirected pecking (Blokhuis, 1989). Problems could arise even in littered systems if the littered area is of insufficient size. Studies in commercial aviaries in Sweden showed that hens used the litter area primarily for foraging, but that most aggression

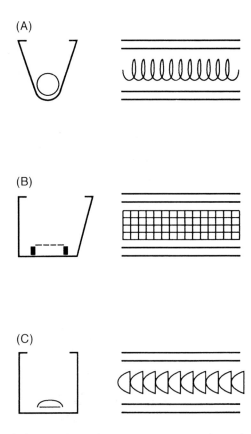

Fig. 4.3. Food trough designs that reduce wastage. (A) Fixed spiral. (B) Fixed grid. (C) Moving chain that both distributes food and minimizes wastage.

and feather pecking also occurred on the litter (Odén *et al.*, 2002). The litter was occupied at high density, particularly in the afternoon, and it was concluded that the litter area was insufficient for the hens to show their preferred spacing during foraging.

Early access to litter may be particularly important; chicks in a commercial aviary raised for the first 2 weeks on slats and then given free access to litter pecked the feathers of other birds more at 5 weeks of age than chicks raised from day 1 on litter (Huber-Eicher and Sebo, 2001a). It may be possible to provide stimuli in cages, such as coloured spots, which increase ground pecking and reduce feather pecking (Braastad, 1990). However, it will probably be most effective to provide loose material, which also allows the birds to dust bathe (see Section 4.12) and results in better feather and foot condition (Robertson *et al.*, 1989).

4.5 Food Intake and Diet Selection

Because production is generally similar across a broad range of housing systems, the amount of food required in different systems depends upon three main factors:

wastage, which is determined primarily by food trough design; energy requirements, which are influenced by ambient temperature, production rate, activity of the birds, feather covering and body weight; and nutrient density of the diet provided. Food intake therefore varies between housing systems, being lower in cages than in non-cage systems (see Section 14.3), which is an important factor contributing to the lower costs of production of caged eggs. Aspects of commercial management, including illumination levels, also influence feed consumption. Hens show a prefer-ence for feeding in bright (200 lux) as opposed to dim (< 1 lux) light, even though the efficiency of feeding is not impaired in dim light (Prescott and Wathes, 2002). Poultry are often kept in dim lighting to reduce feather pecking or activity, so this lack of illumination may depress feeding activity.

Other factors that affect food intake are physical characteristics of the food such as particle size, colour, taste and smell, and the birds' familiarity with these. The effect of these features may be described as the food's palatability. Particle sizes of about 2–3 mm seem to be preferred by both chicks and older fowl (Bessei, 1973; Perry *et al.*, 1976). In one study (Perry *et al.*, 1976), it was found that pullets selected larger particles (>2 mm), which were of cereal origin and high in energy but low in protein, out of a mash diet. As the pullets reached maturity, this preference was less pronounced, possibly reflecting the laying hens' requirement for the higher-protein diet necessary for albumen production. There is also evidence that chickens, given a choice, prefer particulate diets such as pellets to mash (Calet, 1965). The heating process that pellets have undergone may be a factor, because reground pellets are also preferred to mash. However, beak-trimmed hens appear to have difficulty manipulating pellets, and will consume mash instead if given a choice (Deaton *et al.*, 1987). Not only do hens avoid the smallest particles in the diet, there is evidence that if a diet is too finely ground, it can actually have harmful effects – the particles can accumulate as congealed masses in the oral cavity and pharynx, eventually causing infected lesions (Gentle, 1986a).

Under natural conditions, wild birds are faced with an array of different food items that vary widely in nutritional composition; from these, they are capable of selecting a diet that is adequate for all their requirements. Domestic birds (Hughes, 1984) can similarly choose a diet that provides them with all needed nutrients if they are offered a range of different foodstuffs. They do this by initially sampling most potential food items. For example, domestic fowls that are deficient in a particular nutrient such as calcium or sodium show an increase in generalized searching behaviour, pecking at objects that they would not normally investigate (Wood-Gush and Kare, 1966; Hughes and Whitehead, 1979). They continue to consume those items that are palatable and nutritious (Hogan, 1973; Rozin, 1976). Food selection can be very precise; preference tests have shown that the fowl has specific appetites for such essential elements as calcium (Mongin and Saveur, 1979), phosphorus (Holcombe *et al.*, 1976a) and zinc (Hughes and Dewar, 1971), for vitamins such as thiamine (Hughes and Wood-Gush, 1971), and for protein (Holcombe *et al.*, 1976b). Rather surprisingly, there is no evidence that they preferentially select sodium-containing diets, even when they are sodium-deficient (Hughes and Wood-Gush, 1971; Sykes, 1988), perhaps because, under natural conditions, such a deficiency is most unlikely, so the appropriate behaviour has not evolved.

Domestic fowls thus have effective selection mechanisms, and it has been argued (Emmans, 1975) that this ability to choose an appropriate diet can be exploited to increase dietary efficiency under commercial conditions. The more productive individuals in a flock require a diet higher in protein, minerals and vitamins than the less productive ones, so if a complete diet is formulated to support maximum output it provides expensive nutrients that are surplus to the requirements of the remainder of the flock and are therefore wasted. On the other hand, if a cheaper and somewhat less nutritious diet that meets the needs of the average bird is formulated, then the more productive birds will be unable to achieve their potential. This dilemma can be overcome by offering the diet in two (or more) portions, one suitable for maintenance and based on a cereal grain, so that is it relatively low in cost and relatively high in energy, but low in protein, vitamins and minerals. The other complementary portion or 'balancer', which is suitable for production, is expensive, but high in protein, vitamins and minerals. This approach assumes that, when birds are offered two foods, they will select a diet that allows them to produce as well as they would on the better food alone and will also avoid excess nutrient intake.

A number of experiments (reviewed by Hughes, 1984) have been carried out to test this proposition with growing chickens and turkeys and with laying hens. While there is convincing evidence that birds offered a choice between diets can adjust their intake of protein and energy reasonably accurately, and generally grow or lay as well as those given a single complete diet, there is no consistent evidence that this results in improved food utilization efficiency, for reasons that are not immediately clear. The palatability of the different portions of the diet is probably one important factor: wheat, for example, is preferred to barley, oats or rye (Englemann, 1940; Cowan *et al.*, 1978). It is perhaps expecting too much to argue that a modern hybrid strain should be able to select a precisely balanced diet from a number of components, in such a way as to maximize food conversion efficiency and production. These strains have now been maintained for many generations on a single adequate diet so may have lost some of their progenitors' ability to select their own diet. In any case, there would have been little evolutionary pressure under natural conditions to select diets in such a way as to maximize efficiency. Other factors would have been more important, such as avoiding potentially toxic substances, which often taste bitter or unpleasant, one reason why palatability plays an important role in food selection. The influence of factors such as taste, colour, particle size and tactile characteristics of feedstuffs on food selection merits more investigation.

Poultry come to prefer the kind of food to which they are accustomed, and a major change of diet can result in problems. Unless the new diet is similar in texture, colour and probably taste to the previous one, there may follow a reduction in food intake and, therefore, a decrease in growth rate or egg production. The reduction in egg production that results from moving birds to a new diet has been utilized to induce a pause in production, e.g. in the UK where complete deprivation of food is no longer permissible for this purpose. Instead, the hens are switched from a laying mash to a diet of whole grains, usually oats. The daily amount offered is limited to about 30 g and the birds initially find the new diet extremely unpalatable, do not consume even this amount and so rapidly cease laying (Lynn, 1989). The whole-grain diet is also deficient in protein and minerals, and this is also an important factor in helping to terminate egg laying.

4.6 Temporal Pattern of Feeding

Feeding behaviour does not occur at random. It is organized in the short term into bouts or meals and, on a longer-term basis, generally shows a clear diurnal rhythm, being unevenly spread throughout the day. This topic has been well reviewed by Savory (1979).

The sizes of meals and the intervals between them when no feeding occurs have been studied in both domestic fowls and quail (Fig. 4.4). Large meals tend to be preceded by long intervals and, even more so, followed by long intervals. For small meals, the preceding and following intervals tend to be brief. This shows that meal eating is governed by both hunger and satiety mechanisms (Savory, 1979), with signals from the crop, gizzard and duodenum playing an important part (Savory, 1985; Jackson and Duke, 1995).

Young chicks, which are often kept on very long photoperiods such as 23 h light:1 h dark, show little or no photoperiodicity of feeding early in life, even though they hatch with an inherent circadian rhythm of about 25–25.5 h (Aschoff and Meyer-Lohmann, 1954). They then gradually develop a diurnal rhythm of feeding, especially if moved to shorter photoperiods. Laying birds kept on a 14–17 h photoperiod usually show very marked rhythms, with one feeding peak in the morning and a more pronounced one towards the end of the day. Birds do not feed in the dark when the photoperiod is adequate (greater than ~6–8 h), so the morning peak is presumably caused by refilling of the crop, which acts as the food reservoir and which will have emptied overnight. In the evening, birds fill their crops in order to have enough food to last until the next morning. The evening peak is usually less in non-laying birds, but it is greater when a signal of impending darkness, such as a period of artificial dusk, is given (Savory, 1976), as Fig. 4.5 shows. This suggests that the birds are inclined to feed at this time but, in the absence of a suitable cue, have difficulty in anticipating the onset of darkness. The peak may be more obvious in laying birds because they receive internal physiological cues from the process of egg formation. On a 14 h photoperiod, the timing of the evening peak correlates roughly with the start of shell calcification and an increase in calcium requirements, which results in an increase in food intake at the end of the day, but only on days when an egg is being formed (Hughes, 1972; Mongin and Saveur, 1974).

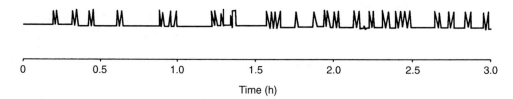

| 0 | 0.5 | 1.0 | 1.5 | 2.0 | 2.5 | 3.0 |

Time (h)

Fig. 4.4. Feeding activity of Japanese quail is organized into separate bouts, separated by intervals when no feeding occurs (after Savory, 1980, with permission from Elsevier).

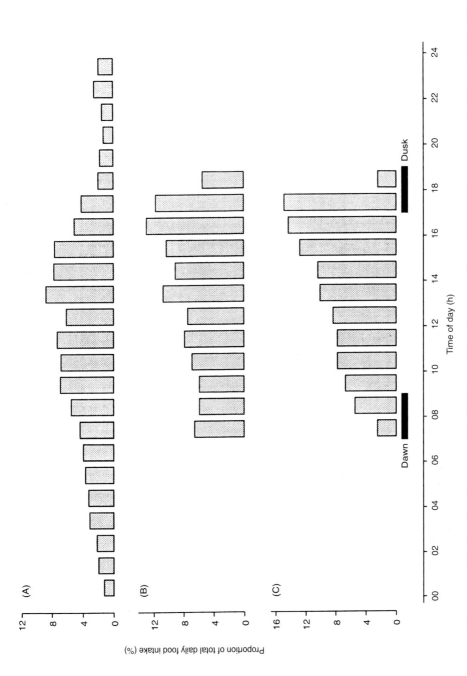

Fig. 4.5. Diurnal patterns of food intake of young cockerels showing how they are influenced by lighting pattern (after Savory, 1976, with permission from Taylor & Francis Ltd). (A) Continuous lighting. (B) Twelve hours of uniform light intensity. (C) Twelve hours of light with a 'dawn' and a 'dusk'.

A number of other factors affect feeding patterns. Laying hens and quail show reduced food intake for about 2 h before oviposition, followed by a compensatory increase afterwards (Woodard and Wilson, 1970). The reduction is not simply due to other behaviour patterns, such as nest investigation and sitting, keeping them from the food trough. Even in cages, where the trough is nearby and the hen may peck at food from time to time, intake decreases similarly, suggesting a specific reduction in feeding motivation during the pre-laying period.

The form and density of the diet also have an influence. Although total food intake was similar in both cases, hens given pellets rather than mash displayed a more pronounced diurnal rhythm (Fujita, 1973). This was because they took longer to consume a given weight of food when it was presented as mash rather than pellets, so feeding time occupied a larger proportion of the day, thus tending to blur the underlying pattern (Fig. 4.6). In the same way, reducing the nutrient density of a diet, for example by diluting it with an indigestible filler such as cellulose powder, also increases total feeding time and again minimizes the diurnal pattern (Savory, 1980). In this case, of course, the weight of food consumed increases to compensate for the dilution. Increasing the time spent feeding by altering the nature of the diet can, in some circumstances, be an advantage, e.g. by reducing the danger of feather pecking and cannibalism. This effect is presumably achieved because of the extent to which pecking activity is directed towards food rather than the plumage of other birds. Dilution of the diet has also been tried as a method of decreasing hunger by increasing gut fill in feed-restricted broiler breeders. While there is some evidence that this can partially reduce feeding motivation (Savory and Lariviere, 2000), the consumption of diluted diets does not consistently result in an appropriate reduction of body weight (Savory *et al.*, 1996).

Birds do not normally feed during the dark period but will do so if the photoperiod is very short, for instance 6 h or less (Morris, 1967). Intermittent lighting patterns are gaining increasing acceptance for commercial rearing, particularly for broiler production, and birds respond by modifying their feeding activity appropriately. Intermittently lit hens housed in cages performed 24% less feeding activity than hens on a 16 h photoperiod, but consumed similar amounts of food. They fed during the 45 min dark periods alternating with 15 min light periods, but did not feed during the longer 8 h dark period, presumably interpreting this as night (Lewis *et al.*, 1987).

4.7 Social Influences on Feeding

As discussed in Chapter 5, social factors can have an important influence on feeding in adults as well as during development. Even in individual cages, domestic hens tend to feed as a group, probably because the sight or sound of one bird feeding triggers feeding in others (Hughes, 1971). Under natural conditions, this would be adaptive, because they would be attracted to join other feeding birds and thus increase their chances of finding food. This propensity towards group feeding does have implications for the provision of feeding space in intensive systems. Ideally, there should be room for all birds to feed at the same time, because at certain times of day a combination of diurnal rhythms and social effects is likely to mean that

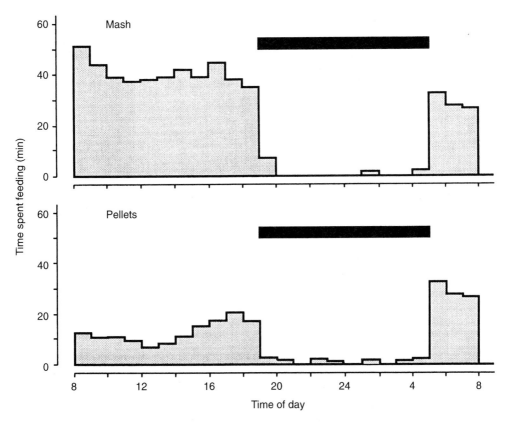

Fig. 4.6. Diurnal variation in time spent eating; the black bar represents darkness. Hens receiving mash spent longer feeding and showed a less pronounced diurnal rhythm than birds receiving pellets (after Fujita, 1973).

most birds are motivated to feed simultaneously. This is illustrated by the finding that, as Fig. 4.7 shows, in wide, shallow cages where each hen had 15 cm of feeder space, birds tended to feed as a group (Hughes and Black, 1976). In contrast, in conventional cages with 10 cm of space, fewer birds fed at a time but feeding activity extended over a greater proportion of the day; those birds that could not feed at peak times had to wait until overall activity had fallen.

When the food troughs of groups of three hens in floor pens were partitioned into three separate feeding areas, the hens spent less time feeding and ate less food than controls with undivided troughs (Huon *et al.*, 1986). In a similar manner, subdividing the feeding space of caged hens by placing dividers along the food trough, thus partitioning it into smaller segments, reduced the time they spent feeding and also the number of agonistic interactions at the food (Preston and Mulder, 1989). This suggests that providing enclosed feeding space may be advantageous in allowing hens to feed with fewer disturbances and influences from other birds than is the case with normal open troughs. The only possible disadvantage is that it also appears to reduce time spent manipulating the food. This may have

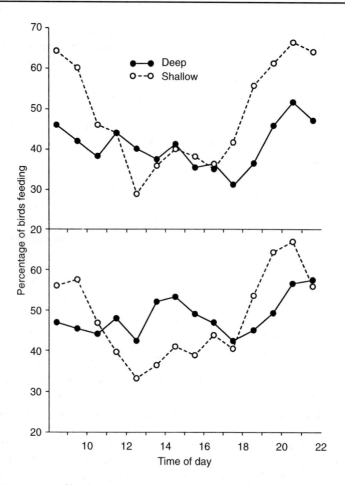

Fig. 4.7. Percentage of birds engaged in feeding activity at hourly intervals throughout the day in either wide, shallow cages, or conventionally narrow, deep cages. The upper graph shows the pattern when there were four birds per cage, the lower graph when there were three. In the shallow cages, more birds fed at the peak times of morning and evening (from Hughes and Black, 1976, with permission from Taylor & Francis Ltd).

implications in that it releases more 'free time', which, if directed towards other birds, could increase the amount of feather pecking.

4.8 Drinking Behaviour

At moderate ambient temperatures, there is a close correlation both hourly and daily between food intake and water consumption (Savory, 1978), and there is consequently a clear circadian pattern of drinking (Wood-Gush, 1959b), with an increase in consumption towards the end of the day because of the evening feeding peak. Adult domestic fowls drink about 150–200 ml of water per day at normal ambient temperatures. This quantity can be consumed in a relatively short time: Gibson *et al.* (1998) found that hens in a covered strawyard spent about 6% of the photoperiod in drinking behaviour. However, this proportion of time can be much longer in cages. Bessei (1986) reported that caged hens spent on average 8 min of each hour (14%) engaged in drinking behaviour.

For laying hens in floor systems, one bell drinker per 100 birds is recommended. This provides about 1.2 cm per hen of drinking space. Alternatively, one nipple drinker or cup may be provided for every ten birds; recommendations for broiler chickens are one per 12 birds, with nipple drinkers preferred to bell drinkers to reduce spillage and hence litter quality problems. In battery cages containing up to six hens, one nipple drinker is generally regarded as sufficient, but the UK Code of Recommendations (Department of Environment, Food and Rural Affairs, 2002a) recommends that each hen have access to two nipples in case one should become ineffective, and this is achieved by placing them at the partition between the two cages. Although the relatively brief time spent at the drinker would imply that competition for space is unlikely, in fact there is evidence that both the amount of drinking space and the way in which the water is presented can influence intake. As the number of hens per nipple drinker is reduced, their water intake goes up. Hearn (1976) found that daily water intake per hen was 165 ml when the number of hens per nipple was ten, 169 ml when there were five, and 182 ml when there were 2.5. Intake increased still further to 213 ml per day when water was supplied in troughs. However, this modest constraint had no effect on egg output.

Because drinking from nipples is not a natural behaviour, birds develop a number of different strategies for obtaining water from them (Hill, 1977). Some peck at the nipple, some hold the plunger up and drink the water as it flows over their beak, some peck at water droplets on the cage structure and some prefer to drink from the drip cups under the nipples. Sometimes pullets are reared from day-old with a particular type of drinker, and then, at point-of-lay, are moved to a different housing system with a different type of drinker. At this age, there is a danger that they may fail to drink from the new drinkers, either because they do not know how to trigger the drinking mechanism or because they do not recognize that the drinker contains water. Even for chicks, dipping the beak into water results in them starting to drink earlier, presumably because they learn to recognize the water more quickly (Yeomans, 1987).

In some litter-based systems, particularly for broilers, problems can arise with wet litter. This may be due to inadequate ventilation or poorly designed drinkers where, at high stocking densities, pushing and jostling between birds causes the drinkers to be tipped and water to spill. Other possible causes are nutritional: a diet that is too high in minerals, especially sodium or potassium, can lead to overdrinking, while high-fat diets have also been implicated. The wet litter can result in high concentrations of atmospheric ammonia, hockburn, necrosis of the feet and breast blisters (Fig. 10.1).

Overdrinking can also cause wet droppings and wet litter in floor systems, and there is evidence that overdrinking can be a stress-related behaviour. Lintern-Moore (1972) observed that a certain proportion of caged laying hens consumed more water than normal hens and also produced particularly wet droppings; she concluded that the cause was behavioural polydipsia or psychogenic overdrinking, a common problem in animals housed in barren environments. Birds that are food deprived may also show increased drinking behaviour (Savory *et al.*, 1992). Food-restricted broiler breeder hens supplied with *ad libitum* water will overdrink and, because of the problems this causes with litter quality, the water supply to breeders is generally limited to only a few hours per day. This could, however, exacerbate the frustration caused by food restriction (Mench, 2002).

4.9 Movement

The word 'movement' (and associated consideration of freedom of movement) combines two aspects of behaviour that can usefully be thought of separately: the relatively small-scale actions involved in actual performance of all behaviour, and larger-scale locomotion. Even in small-scale terms, measurement of the area occupied by hens has shown that conventional battery cages must restrict freedom of movement. A medium hybrid hen, unconstrained and including the tail and other feathers, occupies between 475 and 600 cm^2 when at rest and more if active (Bogner *et al.*, 1979; Freeman, 1983; Dawkins and Hardie, 1989). This area is of course affected by posture, but at space allowances of less than 475 cm^2, medium hybrid hens must frequently overlap or have their feathers compressed. Crowding restricts behaviour, and extreme crowding may also be directly detrimental to welfare (Mench and Keeling, 2001).

No other poultry production system is so restrictive of movement as battery cages. For laying hens, even provision of the 750 cm^2 per bird that will be required in the EU by 2012 would allow 13.3 birds/m^2. There is no single-tier floor system in which it is recommended that birds are stocked as densely as this, and in fact the new EU standards require than hens in non-cage systems be given more than 1100 cm^2 of space each (Chapter 12). Freedom of movement is reflected in the actual number of movements made by birds. One study comparing different systems (Knowles and Broom, 1990) found that hens took an average of 72 steps/h in cages and 208 in a perchery. Wing movements occurred twice per hour and flying 0.4 times/h in the perchery, whereas the latter was completely absent in cages. Another study found similar differences in wing flapping between hens housed in deep litter systems and those in cages (Norgaard-Nielsen, 1990). These differences affect bone strength. Tibia strength is increased by up to 41% and humerus strength by up to 85% in percheries and deep litter systems compared with cages (McLean *et al.*, 1986; Knowles and Broom, 1990; Norgaard-Nielsen, 1990). Bone strength and structure may also be improved in cages simply by adding a perch, although not as much as in alternative systems (Hughes and Appleby, 1989). Weak bones are more likely to be broken both within the system and when birds are removed for slaughter (Knowles and Wilkins, 1999). Up to 30% of caged birds suffer broken bones during catching and transportation, and more during processing, but there are around half as many breakages in birds from free range or percheries as in caged birds (Gregory and Wilkins, 1989; Gregory *et al.*, 1990).

Restriction of movement will also result in the prevention of specific behaviour patterns, because these need more space than standing (Fig. 4.8; see Table 8.1). Such prevention may cause frustration, as discussed later in the chapter, and restriction of

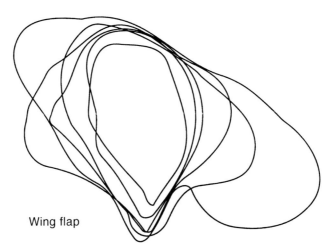

Wing flap

Fig. 4.8. The space used for wing flapping by an unrestricted hen. Successive outlines of the bird were drawn from an overhead video picture, starting with the smallest outline when the bird was standing still (Dawkins and Nicol, 1989). On average, wing flapping used 1876 cm^2 of space.

movement can also have physiological consequences. Birds use postural changes such as erecting their feathers or elevating their wings to dissipate heat, so their ability to thermoregulate by behavioural means will be decreased under crowded conditions.

The question of whether larger-scale locomotion is important to birds prompts contrasting answers. Members of the general public tend to believe that poultry and other animals should be able to move widely. In contrast, there is a widespread opinion among poultry farmers that birds in deep litter or similar systems will not move long distances even to reach facilities such as feeders or nest boxes and that these must therefore be distributed evenly through the house. In fact, there is little evidence for either of these beliefs. Broiler breeder chickens will move readily for nesting or other behaviour (Fig. 6.12). Even broilers, which are commonly regarded as almost immobile in commercial conditions either by nature or because of crowding, actually move about extensively (Fig. 4.9), although this decreases with age as the birds become heavier and develop leg problems (Weeks *et al.*, 2000). However, there is no evidence that such large-scale movements are necessary for the birds if there are adequate facilities close by – in the wild, birds often need to move long distances to locate resources, but this is unnecessary in commercial production systems. Hens and broilers given access to range may make little use of it (Keeling *et al.*, 1988; Weeks *et al.*, 1994). However, many factors could certainly affect this, including how much cover is available to the birds (Dawkins *et al.*, 2003) and how much exposure the birds have to a variable environment during rearing (Grigor *et al.*, 1995a). The quality and variety of resources on the range are probably also important factors with regard to the birds' motivation to use them. For example, broilers are more motivated to move to adjacent areas if those areas contain novel objects or supplemental resources such as preferred dust bathing substrates (Newberry, 1999).

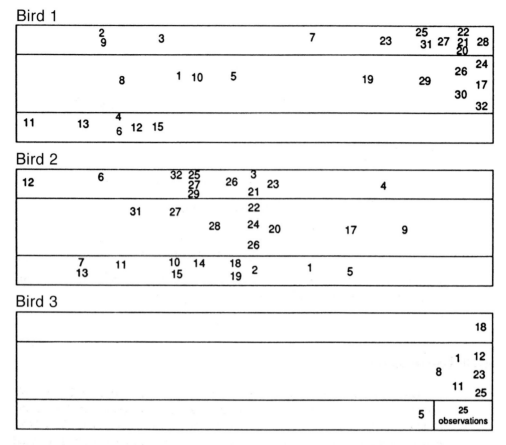

Fig. 4.9. Movements by individual broilers in a flock of over 18,000. Numbers show successive, twice-daily locations from day 27 to day 42. Some moved gradually along the house (bird 1), some moved up and down frequently (bird 2) and some moved very little (bird 3) (from Preston and Murphy, 1989, with permission from Taylor & Francis Ltd).

Regardless, it remains true that restriction of large-scale movement is often associated with restriction of small-scale movement, because of the confounding effect of stocking density. For example, in one study comparing cages with a perchery, hens were active for a similar proportion of time (0.91 and 0.85, respectively) but the mean distance moved was seven times greater in the perchery (McLean *et al.*, 1986). Mobility is directly affected by density: in a deep litter system studied over a range of stocking densities, time spent in locomotion declined at higher densities (Fig. 4.10). However, not all movement is beneficial. When hens from one perchery were examined, some were found to have bones that had been broken, then healed (Gregory *et al.*, 1990). This may have been because they had to jump a large gap to reach the nest boxes and some failed to do so successfully (Broom, 1990). Hens also have difficulty moving between horizontal perches or tiers if the gaps between them are large (Scott and Parker, 1994), a problem that seems to

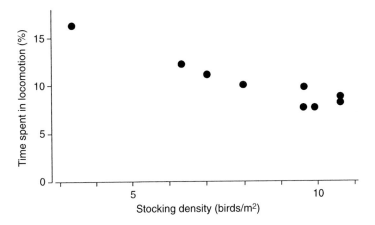

Fig. 4.10. Effect of stocking density on mobility in nine deep litter flocks (from Appleby *et al.*, 1989, with permission from Taylor & Francis Ltd).

be exacerbated if they have been reared without perches (Gunnarsson *et al.*, 2000). Clearly, future designs of percheries and aviaries should take this problem into account.

4.10 Use of Space

The belief among producers that poultry in large houses will not move far probably stems from the idea that they use well-defined home ranges, similar to those in the wild but much smaller. This was put forward by McBride and Foenander (1962), who suggested that birds in large flocks would restrict their movements to small areas in which they could recognize other individuals. In fact, home ranges in large houses are usually either ill defined or maintained by only a few birds. For example, in both deep litter and strawyards, pen area is used unevenly by individuals and by the whole flock, but individual ranges overlap extensively (Gibson *et al.*, 1988; Appleby *et al.*, 1989). In a commercial deep litter house for 4000 broiler breeder chickens, individual ranges averaged 73% of the area (Fig. 5.5). In small pens and in cages, use of space is affected by social rank. Among laying chickens in pens, both males and females of high rank have smaller ranges than those of low rank (van Enckevort, 1965; Pamment *et al.*, 1983), presumably because the latter are avoiding the former while reaching facilities such as feeders. Similarly, in the cages studied by Keeling and Duncan (1989), the top-ranking hen was able to use preferred areas for a greater proportion of the time than other birds. This effect means that there are more constraints on behavioural synchrony (see Section 5.7) for low-ranking birds.

4.11 Comfort Behaviours

Preening and other comfort behaviours, such as wing flapping, feather ruffling and stretching, are important for keeping the plumage well groomed in both natural and artificial conditions. During preening, for example, the feathers are oiled with lipids from the uropygial gland (Fig. 4.11), which helps to maintain good feather condition, and birds will also dislodge and consume parasites living on their skin, such as ticks, while preening (Ostfeld and Lewis, 1999). These behaviours vary between systems in frequency, form, synchrony and, to some extent, also function. This variation is primarily associated with stocking density, because comfort behaviours require a large area for performance (Fig. 4.8; Table 8.1). In hens, they are therefore less frequent in battery cages than in more spacious systems and less frequent in small cages than large ones (Nicol, 1987a,b; Tanaka and Hurnik, 1992). To a lesser extent, they are also constrained by cage height (Nicol, 1987a) and in fact the cage height of 35–40 cm currently required by the EU restricts quite a lot of behaviour. Hence the new EU requirement for enriched cages is that at least 600 cm² of space per hen be 45 cm high. With unrestricted height, nearly 25% of hens' head movements occur above 40 cm. When hens are moved out of small cages, they perform comfort behaviours at an increased frequency, which suggests that constraints on comfort behaviours cause frustration (Nicol, 1987b).

Some comfort behaviours vary in form with space allowance. In particular, preening can be performed in less space at higher stocking densities (Dawkins and Hardie, 1989). However, it is likely to be less efficient, especially as close contact between birds and against the wire will frequently result in feathers being out of place. This may exacerbate the problem of feather pecking, which is on average worse in cages than in other systems (see Sections 5.9 and 5.10). Despite the fact that this behaviour requires less space, preening is also less synchronous in small cages than in large ones, with birds often preening on their own rather than all together (Jenner and Appleby, 1991).

Comfort behaviours have an additional relevance to welfare because they do not always seem to be functional: or rather, they appear in contexts that suggest they have functions in addition to increasing body comfort. Thus birds that are prevented

Fig. 4.11. Birds preen to keep their feathers in good condition. During a preening bout, the bird uses its beak and face to distribute oil from the uropygial gland located at the base of the tail throughout the feathers (photograph courtesy of Cleide Falcone).

from reaching food often preen themselves, but in a slightly faster, more incomplete manner than normal (Duncan, 1970). Such performances of behaviour, apparently in an inappropriate context, are called displacement activities, and are interpreted as indicating frustration.

4.12 Dust Bathing and Water Bathing

The behaviour of bathing in either water (waterfowl) or dust (galliforms and ratites) helps birds to maintain their plumage condition. These behaviours are different from the other comfort behaviours in that to be performed in their complete form they require either loose material or water. Waterfowl raised commercially are rarely given water that is deep enough for them to perform water bathing. The effects of bathing water deprivation have not been well studied.

Dust bathing, however, has been studied in some detail because of welfare concerns about the barren environment of the conventional battery cage. A dust bathing bout begins with the bird pulling loose substrate close to its body. Fluttering movements work this material up into the feathers, where it helps to distribute or remove oily secretions (Fig. 4.12). Although dust bathing thus occurs most often in housing systems with loose material, it can also occur in other housing systems in a 'vacuum' form, in which the bird carries out similar actions on slats or wire, although in longer bouts (Vestergaard *et al.*, 1990). This is sometimes interpreted as indicating high motivation, in which case birds deprived of loose material might suffer frustration. That birds are strongly motivated to dust bathe is suggested by an experiment demonstrating that chicks compensate for an interrupted dust bathing bout by dust bathing more than is typical the next time a substrate is available (Vestergaard *et al.*, 1999). Preference experiments, though, have failed to demonstrate consistent evidence for such strong motivation (e.g. Dawkins and Beardsley, 1986). However, hens deprived of dusting material after having been exposed to it for more than 2 years respond with increased corticosterone levels, suggesting that there is stress associated with dust deprivation for experienced birds (Vestergaard *et al.*, 1997). Experienced hens are also willing to work to gain access to a dusty substrate, even if they have not been deprived of the opportunity to dust bathe (Widowski and Duncan, 2000), indicating that the substrate itself has reinforcing properties for them.

The frequency and length of dust bathing bouts are also influenced by other factors, including photoperiod, ambient temperature, light, the visual stimulus of a dusty substrate, the presence of stale lipids on the feathers, and perhaps the sight of other birds dust bathing (Borchelt and Overmann, 1974; van Liere, 1991; Hogan and von Boxel, 1993; Duncan *et al.*, 1998). Therefore, an alternative explanation for the occurrence of dust bathing as a vacuum behaviour is that it has a low behavioural threshold. For example, it is probable that dust bathing can be triggered simply by the sight of food, which is certainly a loose material. Lindberg and Nicol (1997) found that hens housed in modified cages containing a dust bath still performed most of their dust baths in a vacuum form on the wire floor of the cage, mostly in the area next to the feed trough; they suggested that this sham dust bathing might therefore be an adequate substitute for a 'real' dust bathing bout. To

Fig. 4.12. Many birds dust bathe in dry, friable material to regulate feather lipid levels, and possibly to reduce infestation with mites and other external parasites. This hen is working dust bathing material through her feathers using shaking and ruffling movements of her wings and body (photograph courtesy of Anna Lundberg).

the extent that this is true, birds will not suffer from the absence of loose material for dust bathing. However, dust bathing does function to improve feather structure (Healy and Thomas, 1973) and thus helps to maintain the integrity of the plumage. Lack of loose material in some systems is often considered a welfare problem, and the new EU directive (Commission of the European Communities, 1999) requires that furnished cages contain a littered area. This may be justified partly because dust bathing has physical as well as behavioural effects and partly because loose material is used for other behaviour, such as foraging.

Dust bathing has a strongly diurnal rhythm, being mostly confined to the afternoon. This means that in partly littered systems, the litter may be under-used in the morning but crowded later in the day. In one perchery for laying hens, birds clustered at $24/m^2$ to dust bathe (McLean *et al.*, 1986). This rhythm is used in some commercial systems, including some designs of furnished cages, by allowing hens access to loose material only in the afternoon. Rollaway nest boxes can then be provided without risk of hens laying in the dust bathing areas, because most laying occurs in the morning. In quail, however, laying also occurs in the afternoon, so this option is not available.

Birds readily dust bathe in wood shavings or other floor litter but, if finer material such as sand or peat is available, they use this in preference, probably because finer materials are superior at penetrating the feathers to reach the downy portion of the plumage (van Liere *et al.*, 1990; Shields *et al.*, 2004). This preference is apparent even in young chicks or chicks that only have experience of dust bathing in floor litter (Fig. 4.13).

Fig. 4.13. When given a choice of dust bathing materials, even very young chicks prefer a fine substrate such as sand to coarser substrates such as wood shavings, recycled paper or rice hulls, even though these latter substrates are often used as bedding in commercial poultry houses (Shields *et al.*, 2004). Fine substrates are probably better at penetrating the feathers (photograph courtesy of Sara Shields).

4.13 Rest and Sleep

The main pattern of rest and sleep is set by the light–dark cycle. As with most birds, poultry are generally inactive at night and this diurnal rhythm is strengthened in enclosed houses with a completely dark night, compared with systems with natural lighting and more gradual dawn and dusk. It is weakened, in contrast, in light regimes that use continuous light or simply a dim phase rather than complete darkness, and disrupted even more by intermittent lighting programmes without a well-defined night (Blokhuis, 1983; Coenen *et al.*, 1988). These kinds of lighting regimes are common for rearing broiler chickens. There has been very little consideration of the effects on the birds of such disruption, although one consequence may be abnormal eye development. Chickens that are not exposed to at least 4 h of continuous darkness per day during rearing develop a pronounced flattening of the cornea and damage to the retina, which can impair vision (Li *et al.*, 2000). Chickens reared under continuous light are also more fearful than those provided with a period of darkness each day (Sanotra *et al.*, 2002).

In a normal day–night cycle, other periods of rest occur at intervals through the day, usually with some synchrony between neighbours. This can be readily understood by reference to a farmyard flock or other free-ranging group, in which coordinated behaviour is necessary if the birds are to stay together. Under natural conditions, birds are extremely vulnerable to predation when resting, and resting in a group, especially on an elevated area such as a roost, provides some protection. Even in an enclosed commercial house, chickens show more resting behaviour if there is some form of cover available to them (Cornetto and Estevez, 2001). If there are perches or roosts, chickens and other galliforms usually use them for night-time roosting and also for resting during the day (Fig. 4.14). Among hens kept on wire, this improves the condition of their feet (Fig. 9.1). Fitting a perch in cages also tends to benefit feather condition (Duncan *et al.*, 1992) and increase tibial bone strength

Fig. 4.14. Sleeping (above) and dozing (below) by birds on perches (Blokhuis, 1983).

(Abrahamsson and Tauson, 1993), although because birds rest on their keel while perching they may also develop keel-bone deformities (Abrahamsson and Tauson, 1993). Male broiler chickens tend to use perches much less than female broilers or laying strain birds, probably because their heavy body weight and leg problems make it difficult for them to access and balance upon the perches, but they are nevertheless more likely to use perches for resting when they are housed at higher stocking densities (Petit-Riley and Estevez, 2001).

When perch space is limited, the struggling of hens to get on to perches at dusk is often vigorous, and hens will also push through a weighted door to gain access to a perch at night, suggesting that roosting is a strongly motivated behaviour pattern (Olsson and Keeling, 2002). Perhaps because of this strong motivation, resting birds frequently crowd very closely on to perches, particularly when they are young and not yet of full body size. For all hens to perch once they are fully grown, however, an allowance of about 140 mm of perch per bird (Appleby, 1995) is necessary for most strains, possibly less for light hybrids. If more space is available, birds will, in fact, space more widely than this. This strong tendency for birds to perch can be used to commercial advantage. For example, strain-gauged perches are already used to weigh birds automatically (Fig. 9.5) to obtain flock performance data, and it has been shown that chickens will use water-cooled perches during periods of hot weather to help regulate body temperature, which has the potential to decrease heat stress and thereby decrease mortality and improve carcass quality (Reilly *et al.*, 1991).

5 Living in Groups

5.1 Summary

- Under natural conditions, poultry typically form social groups; one male with several females in chickens, same-sex or mixed-sex groups in turkeys, and mixed-sex groups in ostriches. Quails, ducks and geese tend to form larger aggregations, though pairing may occur during breeding.
- Communication can take place visually through postures and displays, or by calling. Calls serve many functions, including warning, threatening, attracting or signalling food. Physical features on the head and neck can also have signal value.
- Socialization begins in the egg and continues during rearing through imprinting and maternal influence; chicks can also learn some behaviours from one another. Individuals also tend to act synchronously, performing the same behaviour at the same time as other birds, feeding, drinking, resting and dust bathing together.
- Assertion of dominance is by pecking or threatening. Aggression is generally low in small stable groups, partly because the top-ranking bird inhibits it in subordinates, higher in medium sized groups and lower again in very large groups. It is enhanced by disturbance and reduced by dim light. A small proportion of birds at the bottom of the peck order (pariahs) may suffer constant aggression.
- Increased stocking density and group size depress production and affect behaviour. In small groups, introduction of a stranger results in stress, but in large flocks, unless there are partial barriers, birds tend to move over the whole house area and there is probably no individual recognition. Any tendency to maintain a personal space is weak and, in some contexts, such as roosting, totally lacking.
- Feather pecking and cannibalism are major problems in commercial poultry keeping. Feather pecking is worse in barren conditions, in bright light, when

there are nutritional deficiencies, during crowding and when food is quickly eaten. Cannibalism is affected by most of the same conditions, except that it is more common in floor systems than in cages. Both types of pecking are influenced by heritable factors. Beak or bill trimming is effective in reducing damaging pecking, but also raises concerns about welfare because of its potential for causing pain.

5.2 Natural Behaviour

The ancestors of the species that have been domesticated display a variety of different forms of social organization (Mench and Keeling, 2001). Some, such as jungle fowl, live in small relatively stable groups. The most common groupings seen in jungle fowl are of several females with one male, with other males being solitary or in small groups (Collias and Collias, 1996). Each group has a regular roosting site and an area in which it usually forages. The same situation is found in feral domestic fowl, which form distinct social groups, each with a home range (Wood-Gush *et al.*, 1978). Conditions are also similar for small farmyard flocks, and social behaviour in these groups is probably very like that of wild birds.

While wild turkeys may similarly live in small mixed-sex groups during the non-breeding season, all-male or all-female flocks are more common (Schorger, 1966; Latham, 1976). Male flocks are made up of brothers that remain together throughout their lifetime, while female flocks are made up of females from different broods. During the breeding season, turkeys are sometimes found in groupings like those of jungle fowl, with one male and several females, but more commonly male sibling groups stay together and court hens at a place (a lekking ground) where many hens have congregated (see Section 6.2).

Like chickens and turkeys, ostriches live in stable social groups, usually mixed-sex mixed-age family groups, although immature birds from different families may form large aggregations during certain times of the year (Bertram, 1992), and 'adoption' of stray juvenile birds by family groups has also been observed. During the mating season, the most common social unit is a mating pair or a trio composed of one male and two females, but single birds are also seen.

Wild Japanese and bobwhite quail, mallard ducks and greylag geese form less stable social groups than jungle fowl, turkeys and ostriches. Bobwhite quail (Johnsgard, 1973) occupy coveys averaging about a dozen birds of mixed ages and sexes during the winter. The composition of these coveys changes in the spring when males and females pair for breeding, after which new coveys are formed. Japanese quail, mallards and greylag geese migrate to and from breeding and feeding grounds, and group composition changes as a consequence of migration. These birds frequently occur in large aggregations, and so in this sense are adapted to large-scale rearing. However, geese have long-term sexual pairings, so this is one aspect of behaviour that may be disrupted in commercial conditions.

5.3 Social Recognition and Communication

Poultry have excellent colour vision and acute hearing, and communication within and between flocks thus takes place mainly by signals provided by postures, displays and vocalizations. Postures and displays are used to signal threat and submission (Kruijt, 1964; Hale et al., 1969; Wood-Gush, 1971), for example, and particularly elaborate displays are given by all of the wild ancestors of domestic birds during courtship.

Vocalizations given by most species of poultry have been studied to some extent. At least 15 different types of calls are given by Japanese quail (Guyomarc'h, 1967; Potash, 1970), and at least 20 by domestic fowl (Wood-Gush, 1971; Collias, 1987). Turkeys and bobwhites also appear to have a relatively large vocal repertoire (Schorger, 1966; Hale et al., 1969; Johnsgard, 1973). Ostriches and Muscovy ducks, however, are rarely heard to vocalize.

In many cases, the function and causation of calls in domestic birds are not well understood. However, calls can serve a number of different functions, including warning flockmates of approaching predators, decreasing distance between flock-mates, signalling threat or submission, or attracting offspring or flockmates to food. Calls are given in a number of contexts, including when birds are in the presence of stimuli that cause fear or that are positively reinforcing, during social interactions and during mating and egg laying.

Perhaps the most striking vocalizations are the ones that males use in territorial advertisement: the 'crow' call of Japanese quail and fowl, the 'boom' vocalization of ostriches and the 'whistle' of bobwhites. These calls carry great distances and are an effective means of territorial defence in the wild, minimizing the need for direct confrontation between males on neighbouring territories (Collias and Collias, 1996). The crow call of roosters is also individually acoustically distinctive (Siegel et al., 1965; Miller, 1978). Crow characteristics in males are correlated with their comb length (Furlow et al., 1998), one indicator of dominance, and males use crow rates and crow quality to assess the dominance status of other males (Fig. 3.1; Leonard and Horn, 1995).

In addition to vocalizations and displays, features associated with the head and neck are important in some species for both communication and social recognition. In chickens, comb size and hue are influenced by sex hormone levels and are indicators of social status (Guhl and Ortman, 1953). In quail, the head and neck area is important for male–female recognition (Domjan and Nash, 1988). The necks of turkeys are featherless, so colour becomes important. Head and neck coloration can change, and vary from white to red to blue (Schorger, 1966; Hale et al., 1969). Male turkeys have a snood above the beak that is normally flaccid or retracted, as well as the caruncle, an area of spongy tissue on the breast. Both of these become enlarged during aggression and courtship (Hale et al., 1969).

5.4 Socialization

The socialization of poultry begins even before hatching (Rogers, 1995). Calls made by developing chick embryos influence the behaviour of the hen by stimulating her to turn the eggs or to return to the nest to resume incubation, and embryos respond to particular behaviours and vocalizations of the hen with calls that further influence her behaviour. Embryos also influence one another with their calls. Quail and domestic fowl embryos make low-frequency sounds that slow the development of more advanced embryos and 'clicking' vocalizations that accelerate the development of less advanced embryos (Vince, 1970), thus ensuring that the chicks in the brood will hatch at around the same time.

In farmyard flocks, young poultry usually grow up in a group of mixed age and sex. Altricial chicks such as pigeons hatch at an earlier stage of development, and hence require an extended period of parental care prior to leaving the nest. Pigeon parents keep their chicks warm until they are better feathered, and also produce a 'milk' in their crops that is fed to the chicks. In contrast, precocial chicks such as galliforms and waterfowl are mobile from hatching, so it is important that they learn to recognize their mother and their siblings. This occurs through a process called imprinting (Collias, 2000). After hatching, the chicks instinctively follow the first moving objects they see and learn their characteristics. As the mother broods them and helps them to find food, they subsequently learn the advantages of staying with her and with the rest of the brood. Maternal imprinting occurs during a sensitive period of about 2 days after hatching, so in normal commercial conditions, where the mother is absent, this learning process is restricted to learning the features of hatch mates. Sexual imprinting, in which birds learn the characteristics of potential mates, follows at a later age.

Behaviour learned from the mother includes foraging and use of the home range area, but birds can learn to feed on their own (see Section 4.3). In galliforms, perching behaviour is also influenced by the mother. In chickens, the mother broods the chicks on the ground until they are 7 or 8 weeks old, then resumes roosting in trees or bushes, initially low down and later with other adults. The chicks jump up to follow the mother (Wood-Gush *et al.*, 1978) and can subsequently perch on their own. In the single-age groups kept commercially, maternal influencing of behaviour is not possible, and birds may fail to learn appropriate behaviour from each other.

Thus, if medium weight chickens are reared without perches, only some individuals will learn to perch as adults, while others in the same groups fail to do so (Fig. 5.1). The latter birds have difficulty reaching raised drinkers or nest boxes (Appleby *et al.*, 1983). They will also have difficulty accessing the perches in floor housing systems, which contributes to mortality since birds on perches are less likely to be cannibalized (see Section 5.10) by flockmates (Gunnarsson *et al.*, 1999). Similarly, turkey poults may fail to drink and may die of dehydration despite flockmates drinking nearby. Once recognized, however, these problems can usually be alleviated. In the case of perching, providing perches during early rearing usually results in all chicks learning to perch (Appleby *et al.*, 1983, 1988a). To encourage drinking by young turkeys, some chicks may be included in the flock, as the poults can apparently learn drinking from them even if not from each other.

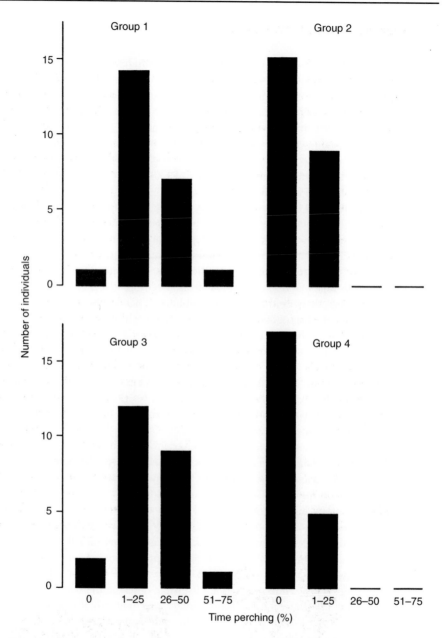

Fig. 5.1. Effect of early experience on perching by adult hens. In this experiment, perching by individual medium hybrids was recorded shortly after they were moved at 20 weeks old to pens with raised nest boxes. Groups 1 and 3 had been reared from 5 weeks with perches, groups 2 and 4 without. Many hens in groups 2 and 4 failed to reach the nests and laid on the floor (from Appleby *et al.*, 1983, with permission from Elsevier).

5.5 Aggression and Dominance

In a farmyard flock, as in a wild group, the first aggression experienced by young birds is probably that received from other members of the flock, when they move too close, or when they are in the way of older birds. Later, particular chicks, ducklings or poults themselves become aggressive to contemporaries. There are also usually certain individuals, perhaps smaller or weaker than others, which are attacked particularly frequently. However, after subsequent hatches, there are younger birds in the flock that they attack in turn. Birds quickly learn that they should avoid others that will obviously beat them, but that they will be able to beat others that are much smaller or weaker. In these circumstances, the most common form of aggression is pecks to the head of the opponent. As the recipient tries to escape, it often receives these pecks on the back of the head. Birds that are more evenly matched are more likely to fight in face-to-face encounters. However, if the group is small enough for members to recognize each other individually, they remember the results of such fights and avoid fighting with others that have beaten them previously.

A relationship between two individuals in which one (the subordinate) avoids confrontation with the other (the dominant) is called social dominance and the set of such relationships in a group is called a dominance hierarchy or peck order. Some individuals high in the hierarchy are able to peck or displace many others, while some individuals low in the hierarchy are displaced frequently (Fig. 5.2). Males and females generally develop separate dominance hierarchies and rarely show aggression towards one another. In a small, stable group, however, even same-sex aggression is usually rare, unless it is provoked by special circumstances such as restricted feeding space, because subordinates avoid dominants whenever possible. Aggression is also low in farmyard flocks because hens suppress aggression amongst their chicks, while roosters suppress aggression amongst the hens.

Commercial rearing conditions are obviously quite different from farmyard conditions, in that the chicks are raised in same-age, and often same-sex, groups. The development of aggressive behaviour in these types of flocks has been studied in some detail. Chickens give aggressive pecks when they are as young as 2 weeks of

Fig. 5.2. Posture and movement of birds show the difference between individuals that are confident or aggressive and those that are not; the bird on the right is showing a submissive posture.

age, although submissive behaviours occur infrequently before 4 weeks. Separate male and female dominance hierarchies are formed between 6 and 10 weeks of age. Chicks initiating aggression at earlier ages have higher initial dominance status (Rushen, 1982), although hierarchies often change when the birds become sexually mature, and male hierarchies are particularly unstable.

Turkeys develop more slowly than chickens, and the onset of aggression and dominance hierarchy formation is correspondingly later (Hale et al., 1969). Aggression begins to be apparent in turkeys at about 3 months of age and increases to a peak at 5 months of age when hierarchies are finally well established. Both males and females form hierarchies, although males fight more vigorously than females.

In established flocks, agonistic interactions are usually subtle and not easily observed, although there can be pecking, chasing and even fighting. This is particularly noticeable among males, and turkey and quail males are especially aggressive to one another, sometimes pecking each other severely enough that the head wounds result in death (Gerken and Mills, 1993; Sherwin and Kelland, 1998). For this reason, turkeys are usually kept under very low light intensities, while quail are generally housed in single-male groups. Young broiler breeder males kept on restricted feed can also be quite aggressive to one another, probably due to feed competition (Shea-Moore et al., 1990).

Among laying hens, the frequency of aggression is generally low in conventional cages, for several reasons. First, birds in a group of four or five know each other well and either have clear dominance relationships or accept equal status. Secondly, if there is a clearly dominant bird, sometimes called a 'despot', this tends to suppress interactions between the others (Hughes and Wood-Gush, 1977; Ylander and Craig, 1980). In addition, cages are too low to allow the birds to raise their heads in a threat, and aggression is generally only provoked by an approaching bird rather than by one in close proximity (Hughes and Wood-Gush, 1977). Aggression is more frequent, though, during times of disturbance such as pre-laying activity, or when the feeding schedule is changed, e.g. when food is withdrawn or returned during an induced moult.

The incidence of aggressive behaviour among laying hens is higher in most alternative systems (Chapter 12) than in cages. The perchery system with litter may be unusual in that, at least in one study, the number of aggressive interactions per bird recorded was lower than in a comparison flock housed in cages (McLean et al., 1986), perhaps because hens were able to withdraw from potential interactions in three dimensions. On the other hand, this was also the rationale behind the design and naming of get-away cages (see Section 12.6) yet ironically these usually had problems with some birds being bullied by others. Housing roosters with the hens in an aviary system was also found to decrease aggression among hens (Odén et al., 1999), as is seen in smaller flocks.

Group size may not be the only factor influencing the generally lower rate of aggression in cages, since aggression is actually higher in smaller than in larger floor-housed flocks (Hughes et al., 1997; Nicol et al., 1999). This may be because birds in larger flocks adopt non-aggressive social strategies for establishing dominance (see Section 5.8). In larger groups, aggression tends to decline as stocking density increases, perhaps because bird movements become restricted and because, as in cages, subordinates have to remain in close proximity to dominants.

It is common in any large group of hens kept moderately intensively that a small number of birds, those mentioned above as being particularly submissive, will be pecked continually by others (McBride, 1958; Gibson *et al.*, 1985). This has been described as the 'peck order effect' (Duncan, 1978a) and such birds as 'runts' (Appleby, 1985) or 'pariahs'. They have heads and combs scarred from pecking, poor body condition and posture, and spend most of the time trying to avoid interaction with others. This often means that they feed very little and they usually stop laying. As such, it is to the advantage of both the birds concerned and the producer if they are identified and removed. When isolated, they will resume feeding and laying as normal. The effect is less common in conventional cages or when hens are kept in small groups in furnished cages (see Section 12.6). It is probably more frequent in larger caged groups, because average individual production declines with group size in conventional cages (Hughes, 1975b). Crowding also plays a role, since when there is limited trough space in cages, the lowest ranked hen in a cage tends to go out of production (Cunningham and van Tienhoven, 1983). Group size has fewer effects on broilers, probably because they are unaggressive and too young to have fully formed a dominance hierarchy (Mench, 1988; Estevez *et al.*, 1997).

If aggression is frequent overall, it can be reduced, along with other activity, by dim lighting. This option is only available in fully enclosed houses; not, for example, in strawyards with partially open sides. For turkeys, providing visual barriers and pecking substrates has also been shown to decrease injurious aggression (Sherwin and Kelland, 1998; Martrenchar *et al.*, 2001). The other main management technique that reduces the effects of aggressive pecking is beak trimming, discussed in Section 5.10.

Aggressive pecks sometimes break the skin on the head or comb. Injured birds should be isolated quickly, both to avoid further injury and because pecking in such circumstances can lead to cannibalism. If they cannot be isolated and the injury is insufficient to justify culling, daubing tar or a proprietary equivalent around the wound is sometimes effective in inhibiting further pecking by other birds.

5.6 Affiliative Behaviour

The main affiliative behaviour shown by poultry is flocking. The tendency to form groups rather than move independently or avoid other members of the species evolved primarily for protection against predators. Even in the absence of predators, birds in large areas generally clump together. This can be easily seen in floor housing systems and is particularly clear in free-range systems. Birds that go out on to pasture from a free-range house generally move as a flock. They often stay near the house (Fig. 5.3), which may be partly due to attraction to the rest of the flock in the house.

A behaviour pattern of more immediate mutual advantage is the habit of pecking food that has adhered to the face of another bird. The bird being pecked remains very still, often with its head back and its eyes closed, allowing the pecking to continue. This probably happens more often at high stocking density where birds are feeding in close proximity. In hens, it may also happen more often in cages than

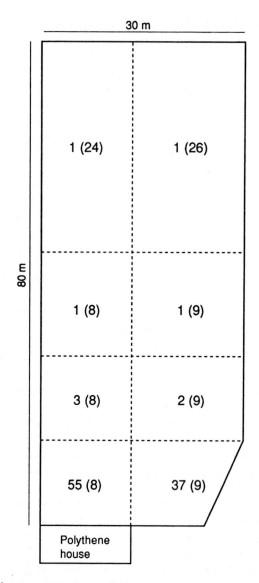

Fig. 5.3. Birds in free range use the available pasture unevenly, usually tending to stay near the house. This diagram shows the percentage use of different parts of a pasture in one study; figures in parentheses show the percentage of total area (Keeling *et al.*, 1988).

in other systems because of the lack of suitable objects on which birds can clean their own faces by wiping their beaks. There is a possibility that this could produce a disadvantage despite the obvious advantage. Such behaviour may be a predispos- ing factor to feather pecking or cannibalism, because birds being pecked in those cases frequently also freeze, rather than trying to escape.

5.7 Social Learning and Behavioural Synchrony

Poultry are keen observers of one another's behaviour and can learn new behaviours by observing their flockmates. For example, chickens that have an opportunity to observe a trained hen peck a key to obtain food learn this task much more quickly than chickens that do not have this experience (Johnson *et al.*, 1986). Birds can also learn from observing humans. In one study, semi-feral greylag goslings that were imprinted on humans learned to open a box to get at a food reward after observing a human 'tutor' (Fritz *et al.*, 2000).

Even when birds do not learn behaviour from each other, however, they readily copy behaviour in such a way that many activities are performed synchronously in a group. This is particularly clear with feeding, and in all husbandry systems it is common for birds to feed together rather than at different times. Some of this synchrony can be accounted for by factors that act separately on all birds, for example the light regime, which produces peaks of feeding at dawn and dusk quite apart from social influences (Section 4.6). In addition, however, bouts of feeding through the day are more synchronized between birds than would be expected at random (Hughes, 1971). This pattern is constrained for hens in cages that provide a limited amount of trough space per hen, not so much because of the feeding space itself as because the cage is too narrow for all hens to stand side by side (Fig. 5.4). This results in hens contending to feed simultaneously. This problem is avoided in shallow, wider cages and also in large flocks fed *ad libitum* where birds can easily feed separately. It is worse in breeding flocks on restricted food, in which all individuals react to the arrival of a new delivery of food. Because the birds have more freedom to orient themselves around round feeders, these allow more birds to feed simultaneously than linear troughs even when the same actual feeding space is provided.

Synchrony in resting is obviously also influenced by the light regime, but also occurs during the daytime, interspersed with bouts of feeding. Synchronous roosting is advantageous in cold conditions, when roosting in contact conserves body heat. Again, roosting side by side is constrained for birds in small cages, but is not a problem in larger cages or floor systems. In systems such as deep litter and aviaries, 'rafts' made up of hens in close body contact form at night. This happens even when lights go off abruptly. However, roosting behaviour is facilitated in houses with a gradual dusk, or where lights-off is preceded by a dim light period. This is particularly important when roosting above ground level, on perches or on a platform, is to be encouraged, e.g. in houses with partially slatted floors that allow for the accumulation of droppings produced during the night.

Other behaviour patterns, such as drinking, preening and dust bathing, also tend to be performed synchronously (Mench *et al.*, 1986; Webster and Hurnik, 1994). As with feeding, synchrony is greater in the small groups of laying hens housed in cages, but may also be restricted in cages. Preening, for example, requires considerably more space than most caged laying hens are given (Table 8.1), so not all birds in a cage can preen simultaneously.

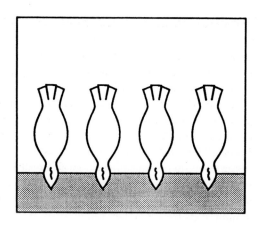

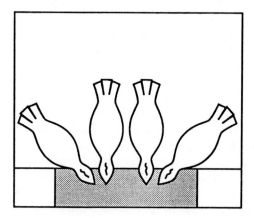

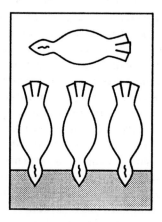

Fig. 5.4. Examples of the orientation of feeding birds when given adequate cage and trough width (top), adequate cage width but restricted trough width (middle) and restricted cage and trough width (bottom) (Hughes, 1983).

5.8 Stocking Density, Group Size and Spacing Behaviour

Stocking density and group size can affect both production and behaviour. In conventional cages, higher stocking densities and larger group sizes are associated with decreased egg production, higher mortality, more feather pecking and cannibalism and increased fearfulness (Adams and Craig, 1985; Mench and Keeling, 2001). In general, small group size is advantageous. For example, in cages for laying hens, small groups show higher production levels compared with larger unit sizes (Hughes, 1975b). There is also some evidence that in cages, stress decreases linearly with decreasing group size (Mashaly *et al.*, 1984; Roush *et al.*, 1984). Furnished cages that retain small group sizes may have similar advantages (Appleby, 1998). However, hens do not necessarily prefer small group sizes unless adequate space is also provided (Lindberg and Nicol, 1996).

When birds are kept in large groups, group size has other effects on behaviour that are independent from those of stocking density: (i) at the same density, a larger group has a larger absolute space in which to move; (ii) in large groups, there are more individuals with which to interact and frequency of social interactions increases with group size; and (iii) birds in large groups will have more difficulty learning to recognize their flockmates individually.

In large pens or houses, birds do not use the area evenly (Section 4.10), but they generally move about sufficiently to suggest that they encounter all other members of the flock. This has been found for laying hens in strawyards (Gibson *et al.*, 1988) and on deep litter (Appleby *et al.*, 1989) and for broiler breeders on deep litter (Fig. 5.5). Individual recognition is not possible in these conditions. Although it is theoretically possible that dominance hierarchies can exist without individual recognition (Wood-Gush, 1971; Barnard and Burk, 1979), it is generally thought unlikely that birds in large flocks could form a hierarchy. Thus, birds in large commercial flocks may be in a constant state of trying to establish a hierarchy but never achieving it.

It is not known whether birds become used to these continual encounters with unfamiliar individuals but, in small groups of chickens, contact with strangers results in increased heart rate (Candland *et al.*, 1969), increased aggression (Craig *et al.*, 1969) and growth of the adrenal glands (Siegel and Siegel, 1961), which are indicators of stress. In some houses, though, subgroups may form within pens. For example, in one commercial aviary, hens that roosted near the ends of the aviary were consistent in using that same area during the day, although hens that roosted in the middle of the aviary distributed themselves more randomly (Odén *et al.*, 2000). In an experimental aviary with nest boxes in the centre of the house (Hill, 1983), the hens rarely moved over these even though they were not prevented from doing so. House features can thus be important in facilitating subgroup formation. Hens in large groups may also move away from a mechanism for social relationships based on individual recognition to one where dominance relationships are determined by physical factors such as body and comb size (Pagel and Dawkins, 1997).

Within groups, the spacing of birds varies with the activity that they are performing. While they will roost in body contact and preen quite close together, they are usually more spread out (subject to the constraints of their housing) while foraging for food on pasture or in litter (Keeling and Duncan, 1991). This is partly a

Hen 1   Hen 2   Hen 3

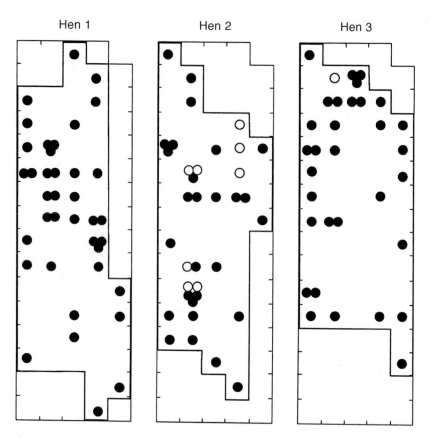

Fig. 5.5. Movement of broiler breeders in a deep litter house, 46 m × 15 m. Birds were chosen at random from a flock of nearly 4000, and their locations recorded on 52 days over 34 weeks; their ranges were all more than half the area of the house. Open circles represent nesting, and filled circles are other records (Appleby *et al.*, 1985).

mechanical effect, because pecking and scratching result in a certain separation. However, it is also partly a social effect, as one bird approaching another sometimes provokes aggression or retreat. This has sometimes been described as defence of a 'personal space' by each bird, but this is not as clear-cut as in some other animals, e.g. any such tendency is clearly absent during roosting. Nevertheless, while birds do clump into flocks, as described in the previous section, they react to high stocking density by spacing more evenly than random. Thus, in cages for laying hens, where close proximity is enforced, birds attempt to stay further apart than random. This even occurred in experimental pens for three hens that provided 1400 cm^2 each (Keeling and Duncan, 1989). It is not known how this sort of spacing is affected by use of three dimensions in housing. For example, in a perchery or aviary, hens are in close proximity above and below each other as well as horizontally, and it is possible that this also causes some stress. While vertical movement may be common in the wild, as when jungle fowl or pheasants roost in trees or bushes, this would not be at densities comparable with those in commercial systems.

5.9 Feather Pecking

The problem of feather loss in laying chickens, and to a lesser extent in turkeys and pheasants, has received more attention in consideration of different housing systems than almost any other issue. The main reason for this attention was probably initially aesthetic: birds with extensive feather loss are unattractive. However, such loss (apart from moulting) also indicates major behavioural or physiological departures from natural conditions and increases the danger of exposed skin being injured. The main cause of feather loss in all systems is not physiological change or abrasion, but feather pecking (Hughes, 1985). It is painful for a bird to have a feather pulled out (Gentle and Hunter, 1991), and feather loss can also be an economic problem, since birds with few feathers lose heat faster and therefore consume more feed.

There are two kinds of feather pecking: gentle pecking that results in little damage (sometimes called allopecking or allopreening), and severe feather pecking that results in feather damage or loss. Feather pecking is different from aggressive pecking, both in character and in effect. The movements involved are not rapid and violent, as in aggression, but instead deliberate and similar to feeding movements (Wennrich, 1975). In more severe forms, the feathers are grasped and then pulled. Pecking is often directed at feathers that are damaged or distinctive, or which are out of line (McAdie and Keeling, 2000). Juveniles most commonly peck the small oilier feathers near the tail and preen gland (Savory and Mann, 1997), while in older birds, at least initially, the back is pecked, perhaps because feathers that are out of line in more accessible places are quickly preened. Damage can then progress to the tail and even to the whole body (Fig. 5.6). Feather loss from aggressive pecking, on the other hand, is usually confined to the head.

Feathers that have been removed are sometimes eaten. Feather eating is probably an abnormal feeding behaviour and can lead to the formation of feather balls in the crop and subsequent weight loss and crop impaction (Morishita *et al.*, 1999). Since it is common for floor-housed chickens to eat feathers that they find on the ground, it may be that feather pecking develops if insufficient numbers of loose feathers are available to be consumed so that instead feather-eaters direct their pecking towards other birds (McKeegan and Savory, 1999).

While many individuals in a group may show feather pecking behaviour, some individuals are particularly liable to peck others at high rates and more severely, while some individuals are prone to being pecked (Vestergaard *et al.*, 1993; Wechsler *et al.*, 1998). There are differences between commercial strains in the incidence of feather pecking, and feather pecking behaviour is moderately to highly heritable (Kjaer and Mench, 2003). There have been experimental studies of selection against feather pecking, but more research is needed to show whether this is practical commercially.

There are also major environmental influences on the behaviour. Predisposing factors identified in a recent survey of alternative housing systems in the UK were dietary changes, low temperature, high lighting levels during inspection, the use of bell drinkers, lack of use of the outdoor area and absence of loose litter at the end of lay (Green *et al.*, 2000). As the last two factors suggest, feather pecking is worse in barren conditions, presumably because the availability of other, varied stimuli for

Fig. 5.6. An extreme case of feather loss. While some feather loss is caused by abrasion, most is due to feather pecking.

pecking is then reduced (Blokhuis, 1989). It is therefore often a major problem in cages, reflected in the fact that worse feather loss has often been recorded in cages than in other systems (McLean *et al.*, 1986; Appleby *et al.*, 1988b). For example, loss in a perchery was 2.7 compared with 4.4 in cages (on a scale of 0–20; McLean *et al.*, 1986). In a study of free-range hens, feather damage was also less than in cages. Scores over 4 years were 1.2–1.5 compared with 1.8–3.5 in cages (on a scale of 1–5; Hughes and Dun, 1986).

Nevertheless, feather pecking also continues to be a significant problem in alternative systems. An examination of feather pecking by hens kept in 25 commercial aviaries and deep litter houses showed that nearly 80% of flocks displayed high frequencies of feather pecking by the time the birds were 14 weeks old (Huber-Eicher and Sebo, 2001b). Feather pecking may happen if birds are housed such that some birds defecate on others. In these circumstances, hens peck at soiled feathers and pecking may then spread, either because the birds are attracted to pecking damaged feathers (McAdie and Keeling, 2000) or because the birds copy one another's behaviour (Zeltner *et al.*, 2000). In one aviary system that had such an arrangement of tiers, very severe loss was recorded in the first few flocks housed (Hill, 1983). However, outbreaks of feather pecking are not confined to systems where this is a possibility. High stocking densities may be contributing factors in such cases (Fig. 5.7) and group size also has an influence (Hughes and Duncan, 1972; Bilĉik and Keeling, 1999), with feather damage being more extensive in larger groups.

As already pointed out, feather pecking is similar to food pecking (Wennrich, 1975). It can be triggered by nutritional deficiencies (Wahlström *et al.*, 1998) and is

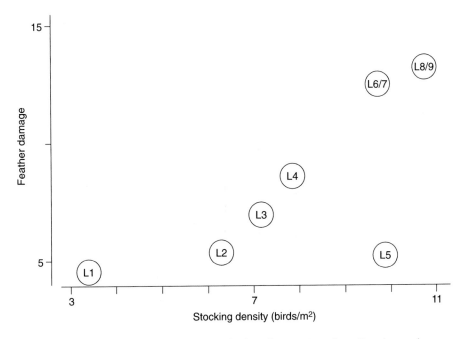

Fig. 5.7. The effect of stocking density on feather damage in a deep litter house for laying hens. Damage was scored at the end of lay on a scale from 0 (no damage) to 20 (denuded). Points represent different flocks and, despite the low damage in flock 5, the effect was statistically significant (from Appleby *et al.*, 1988b, with permission from Taylor & Francis Ltd).

exacerbated by some feeding methods. Poultry can obtain sufficient food much more quickly in domestic conditions than in feral conditions, where birds spend 50% or more of their time in feeding (Savory *et al.*, 1978). Some of the remaining time may be spent in feather pecking. However, where pasture, litter or other loose material is available, birds will forage even if this yields little food (Hughes and Dun, 1986; Gibson *et al.*, 1988), and this probably contributes to the difference between cages and other systems. Experimental evidence suggests that feather pecking can be decreased by providing litter or other appropriate foraging materials (Blokhuis and van der Haar, 1989; Huber-Eicher and Wechsler, 1998; Johnsen *et al.*, 1998), although the beneficial effects of rearing with litter are not as clear under commercial conditions (Gunnarsson *et al.*, 1999). Another major feature of feeding regimes is whether pellets or mash are supplied. This used to be determined by the method of automatic food distribution in a particular system. However, food in the form of pellets can be eaten faster than mash, and this also encourages feather pecking, at least when foraging materials are unavailable (Aerni *et al.*, 2000). For this reason, the use of pellets for laying hens is now rare.

Making substrates available for dust bathing can decrease feather damage, although there is controversy as to whether or not this effect is due to some relationship between dust bathing and feather pecking (Nørgaard-Nielsen, 1997; Johnsen *et al.*, 1998). Providing perches in floor housing systems for hens can decrease feather pecking, although high perches are more effective than low perches since the birds most likely to be feather-pecked are those on or near the floor (Wechsler and Huber-Eicher, 1998).

As with aggression and all other activity, feather pecking can be reduced in enclosed houses by lower light intensity. However, light is usually kept at the lowest level compatible with production in any case, so this is available only as a short-term measure. In cages for laying hens, it is sometimes possible to identify and remove the birds responsible, because, in a small group, such birds will have much better plumage than the others that they have pecked. They cannot be regrouped, however, because if they are they will resume pecking each other. Housing them singly is generally impractical, so the main option is to trim the beaks of these specific culprits. Beak trimming (see next section) does reduce the effect of feather pecking (Hughes and Michie, 1982), at least in the short term. In the long term, its effect is variable. In a study of a deep litter system, feather loss was reduced by beak trimming, while there was no corresponding effect of beak trimming in cages (Appleby *et al.*, 1988b).

5.10 Cannibalism

Cannibalism, which involves the pecking and tearing of the skin and underlying tissues of another bird, occurs on occasion among hens, turkeys, pheasants, quails and ducks. It sometimes follows on from feather pecking, for example when exposed skin is injured, but it more often arises independently. In hens, a common form of cannibalism is cloacal cannibalism, or vent pecking. One situation in which this starts is when a hen has just laid an egg and the cloaca is still partly everted. Other hens peck at the soft, red vent area. If the skin is broken, here or elsewhere on the body, other birds then join in pecking, because these species of birds are attracted to blood. Further pecking and consumption of flesh then frequently result in death. As with feather pecking, specific individuals are likely to show cannibalistic behaviour (Keeling, 1994), although birds that are cannibals are not necessarily the same birds as those that show feather pecking behaviour.

The aspect of cannibalism that is least understood is that the pecked bird often makes surprisingly little effort to escape, despite the fact that it is likely to be in severe pain. Sometimes this may be because the pecked bird is a low-ranking individual, as described above, which has been pecked aggressively so often that it has learned that it cannot escape, and so freezes. Similarly, in a small cage or at high stocking density, birds may learn that they cannot avoid being pecked. One other possible explanation is that this freezing is related to the behaviour that occurs when one bird is having food pecked from its face by another (see Section 5.6). This would suggest that a bird is misled by circumstances (such as, perhaps, high stocking density) into performing an inappropriate and fatal behaviour pattern. In any event, it is evident that cannibalism is a major problem for both the animals concerned and the producer.

The same factors that result in higher levels of feather pecking also result in higher levels of cloacal cannibalism, but flocks do not necessarily experience both problems at the same time. There can be serious outbreaks of cannibalism in caged flocks (Tablante *et al.*, 2000), but cannibalism is more common in non-cage systems.

This may be partly due to crowding or social disturbance: in one deep litter house, cannibalism began on a crowded, slatted area when some hens escaped from one pen into another (Appleby *et al.*, 1989). Cannibalism might also spread because of social learning. In an experimental study where hens were given membranes filled with blood, birds that observed other hens pecking the membrane and consuming the blood imitated this behaviour (Cloutier *et al.*, 2002). It will also be clear from the description of cloacal cannibalism above that there are particular predisposing factors, such as designs of nest boxes that encourage birds to face inwards, exposing their vent area to other birds after oviposition, and inadequate numbers of nest boxes for a flock of birds, resulting in hens having to lay on the floor. There is also a genetic component to cannibalism, since it has been shown that the incidence of cannibalism can be lowered through group-based selection for survival in caged hens (Muir and Craig, 1998).

The single most important environmental factor, however, is group size. Outbreaks of cannibalism are unpredictable in nature, occurring in some flocks but not in others. When they do occur, mortality can be disturbingly high. Losses of up to 13% of a flock of laying hens have been reported in an aviary (Hill, 1986) and of up to 15% in both a strawyard (Gibson *et al.*, 1988) and a free-range system (Keeling *et al.*, 1988). An outbreak of cannibalism can occur at any stage of the laying cycle. In the free-range flock cited, cannibalism became severe after 11 months of lay, with most losses during the final 8 weeks (Keeling *et al.*, 1988).

Reduction of light intensity, if this is possible, can help to stop an outbreak of cannibalism. One other technique that can be combined with this is to introduce red lighting. This makes it more difficult for birds to see blood or wounds, while still allowing them to feed and perform other behaviour. Neither of these methods is available in systems open to daylight, including strawyards. Birds that are behaving as cannibals are sometimes seen with blood on their beaks, in which case they can be removed. This is more often possible in cages than in other systems, because a cannibalized bird in a cage only has a small number of flockmates.

Several methods are used to reduce feather pecking and cannibalism (Fig. 5.8). The birds (especially gamebirds) may be fitted with 'spectacles' or rings that are inserted through the nasal septum. Although the effects of these on welfare have not been studied, spectacles reduce the bird's forward vision and rings prevent the bird's beak from closing properly, both of which raise concerns. More commonly, birds are beak or bill trimmed. In chickens, this procedure usually results in removal of between a third and a half of both the upper and lower mandibles. This involves not just the horny beak but the underlying tissue as well. Use of a special cutting tool with a heated blade to cauterize the bleeding is common (Fig. 5.9), although newer trimmers that use an infrared beam to cut a hole in the beak, causing the beak tip to drop off several days later, are increasingly used, especially for broiler breeders. Beak trimming is done shortly after hatching, usually within 10 days, although laying pullets may be re-trimmed a second time at 5–8 weeks of age if there is re-growth of the beak. Beak trimming probably has two effects on pecking. First, trimming reduces the sharpness of the beak and the accuracy with which the bird can peck. Pecking frequency actually increases; for example, while feeding, more pecks are needed to achieve the same intake (Gentle *et al.*, 1982). Aggressive birds also peck subordinates more often, because the latter react less. However, the

Fig. 5.8. Two hens that were beak trimmed when young, in this case by removing part of the upper mandible. (A) Partial re-growth of the mandible has occurred, producing a beak that looks fairly normal, but in which nerves may be abnormal. (B) Severe beak trimming has resulted in malformation of the upper mandible and overgrowth of the lower mandible.

effect of pecks on the feathers of other birds, from either aggression or feather pecking, is reduced. Secondly, it is likely that hens with trimmed beaks do not peck so strongly because to do so is painful. In fact, studies of the nerves indicate that trimming may result in long-term pain (Gentle, 1986b), although the effects on behaviour and physiology are much less severe when birds are first trimmed at hatch

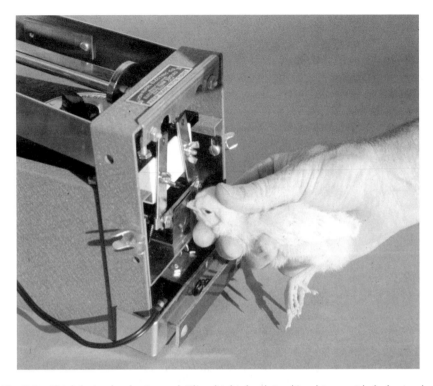

Fig. 5.9. Chick being beak trimmed. The chick's beak is placed in a guide hole sized appropriately with respect to the amount of beak to be trimmed. The machine then cuts and cauterizes the beak. There are also newer designs of trimmers that use an infrared beam to make a small hole in the beak tip, which then falls off several days later (photograph courtesy of Ralph Ernst).

or shortly thereafter (Gentle *et al.*, 1997). Thus, although the main reason for beak trimming is to prevent cannibalism, a welfare problem, the procedure is subject to criticism on welfare grounds.

Beak trimming is used as a preventative measure prior to housing in all systems, and it is usually effective in reducing the likelihood of cannibalism. In a comparison of deep litter with cages, outbreaks of cannibalism occurred in both systems in flocks that had not been beak trimmed, but not in flocks trimmed at 1 day old (Appleby *et al.*, 1988b). However, in some countries, beak trimming has been banned. In the UK it was, until recently, recommended that it be carried out only as a last resort (Ministry of Agriculture, Fisheries and Food, 1987a) because of the criticisms discussed above. The use of beak trimming only as a therapeutic method is rare, however. It was adopted in a study of a perchery, which was divided into pens housing about 120 birds. The house was stocked with birds which had not been beak trimmed; beak trimming was carried out later on the birds in any given pen only if cannibalism occurred in that pen. As it turned out, beak trimming was necessary in five of the six pens in 1 year, but in no pens in the following year (Michie and Wilson, 1985). This approach would be impracticable in a large flock and is no longer recommended in the UK (Department of Environment, Food and

Rural Affairs, 2002a). In welfare terms, current systems housing large flocks of laying hens have two major, alternative problems: the risk of outbreaks of cannibalism and the effects on the birds of preventative beak trimming. These problems must be weighed against the welfare advantages that such systems offer.

6 Reproduction

6.1 Summary

- Ancestors of domestic poultry have varied mating systems. Males show courting behaviour while females select males based on displays and physical features. Nests are usually simple hollows and females generally incubate, though in some species both parents take part.
- Where natural mating is required, mixed-sex rearing is important for sexual imprinting. The presence of males may also stimulate female development. In small mixed flocks, high-ranking males mate most frequently, but rank has less effect in large flocks.
- Males mate about five times per day, while females mate 0.5 times. Fertility is high even with infrequent mating, because viable sperm are stored in the oviduct. Artificial insemination is the rule in turkeys, because broad-breasted strains cannot mate naturally.
- In many broiler breeder flocks, fertility and hatchability decline with age, because of male obesity and male aggression towards females. This may be because of a breakdown in normal courting behaviour, perhaps because of inappropriate genetic selection.
- Egg-laying behaviour is under tight genetic control; certain aspects have been altered by selection for egg production and by inadvertent selection against floor laying.
- Ovulation occurs soon after dawn and starts a sequence that triggers pre-laying behaviour and oviposition 24 h later. Disturbance around this time can disrupt the process, so in floor systems it is important to provide sufficient and adequate nest sites.
- Nest boxes may be poorly used unless birds have opportunities to perch when young. In broiler strains, restricted mobility may also be a factor. Placing pullets

into nests also increases nest box use. Enclosed nests with flat floors and litter are the most preferred, but some types with a gently sloping, artificial turf floor are also acceptable.

- Pre-laying behaviour is normal in small groups with littered nest boxes, but is often abnormal in cages or where nests are wire rollaway. There are strain differences, and this behaviour can be modified by genetic selection.
- Disturbance during shell formation can cause mis-shapen eggs and just prior to oviposition can lead to egg retention and calcium dusting. Abnormal behaviour during oviposition can result in cracked eggs and egg eating.
- Broodiness is a normal behaviour but is undesirable under commercial conditions; its incidence is low because of selection and because eggs are removed from nest boxes.

6.2 Natural Mating Behaviour

Birds display a wide variety of mating systems, and all of these systems are evident in the natural mating behaviours of the ancestors and feral counterparts of domestic poultry species. Mating may be promiscuous (polygynous, polyandrous, or both) or monogamous. Monogamous pair bonds may last for only one season, or be maintained over several breeding seasons. Males may set up territories to which females are attracted for mating during the breeding season, or they may associate with a harem of females year-round and maintain a territory in which those females remain during the breeding season. Alternatively, males and females may congregate during the breeding season at special breeding grounds where the females select males with which to mate, a system called 'lek' mating.

Jungle fowl and domestic fowl have a harem polygynous mating system, with a dominant male maintaining a territory and monopolizing mating in the group of females living in that territory throughout the mating season (Collias and Collias, 1996). Other, socially subordinate, males are tolerated near harems, or they may be solitary or form separate groups. Jungle fowl living in dense vegetation seem to form small harems containing up to four hens (Collias and Collias, 1967), but in more open conditions they form larger groups (Collias *et al.*, 1966) and both hens and cocks in adjacent flocks may periodically move from one flock to another. Studies of feral fowl have described harems of 4–12 hens (McBride *et al.*, 1969), but these must also depend on precise local conditions.

Under some conditions, turkeys also form harems. As with fowl, these harem groups usually comprise a single, older male and about 4–6 females (Schorger, 1966). More commonly, however, wild turkeys mate in leks. Male turkeys normally live in all-male groups comprised of siblings (Chapter 5). As the breeding season approaches, these sibling groups come together at a lekking ground where the females have congregated (Watts and Stokes, 1971). The sibling males together perform displays to court the females. The dominant male in the sibling group usually secures most matings during the height of the breeding season, although other males in the group may mate later.

Ratites normally live in family groups or large aggregations but, during the breeding season, males establish a territory in which they build a nest. The mating system in ratites combines short-term pair bonding, polygyny and sequential polyandry. In ostriches, for example, a primary pair comprising the resident male and a so-called 'major' female is established in each territory and this pair incubates the eggs deposited in the nest in that territory and rears the chicks (Bertram, 1992). However, females also visit (and lay their eggs in) the territories of several different males and males may mate with several females on their territory during a single breeding season.

Bobwhite quail are usually seasonally monogamous in the wild, with males and females leaving their coveys to pair for breeding in the spring (Johnsgard, 1973); unpaired males may set up adjacent territories. Ducks, geese and Japanese quail are more variable in their mating behaviour. Like bobwhites, ducks and Japanese quail are sometimes seasonally monogamous, but both will also mate promiscuously (Mills *et al.*, 1997b). Geese maintain long-term pair bonds in the wild, but will mate promiscuously under domestication. Although wild mallards form pair bonds that can be maintained for more than one breeding season, the males do court other females and also force copulations on unwilling females. Under crowded conditions, female mallards can be severely harassed and even injured or killed by males attempting to mate with them (McKinney, 1975).

Even in apparently promiscuous mating systems, birds (and especially females) do not mate randomly but are selective in choosing potential mates. Mate selection has been studied in most detail in jungle fowl. Females use a suite of physical characteristics to assess an unfamiliar male prior to approach and copulation, including his comb colour, eye colour, spur length and comb size (Zuk *et al.*, 1992, 1995). Comb size is one of the most important cues used by females and, since males infected with parasites have smaller combs, this may be one way in which females can assess a male's fitness (Zuk *et al.*, 1990). Similarly, female turkeys prefer males with longer snoods and wider skullcaps, features that are correlated with lower parasite loads in wild males (Buchholz, 1995). The rate and nature of courtship displays by males seem to be relatively unimportant for female mate choice in jungle fowl, at least when the hen is familiar with the roosters in the flock (Zuk *et al.*, 1990), although domestic hens may refuse to crouch for males that do not perform courtship displays (see Section 6.6).

Once potential mates are selected, courtship consists of a chain of stimulus–response patterns (Fig. 6.1) between male and female (Fischer, 1975). The male is usually the obvious initiator of the courtship sequence, although females can encourage courtship by approaching or maintaining proximity towards certain males (e.g. in fowl and mallards), by congregating at a lekking ground (e.g. turkey hens) or by engaging in pre-nuptial displays prior to the onset of the breeding season (e.g. ostriches). Male courtship displays are generally elaborate, involving vocalizations and noises, conspicuous postures, spreading of the feathers such that the male looks larger and his plumage characteristics are emphasized, and sometimes colour changes or enlargement of certain body features, such as the snood of male turkeys.

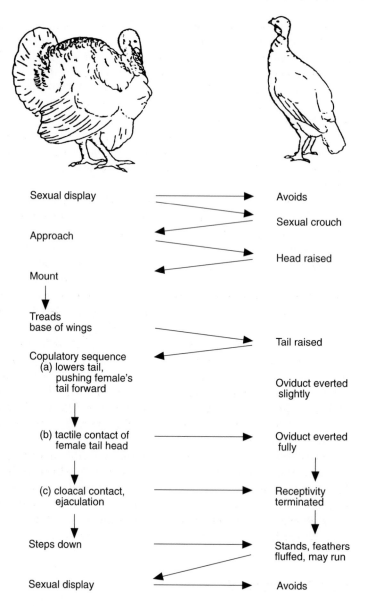

Fig. 6.1. A schematic illustration of the sequence of reproductive behaviour in turkeys (from Hale *et al.*, 1969).

6.3 Natural Nesting and Incubation

Like mating, patterns of nesting and incubation vary among the ancestors of domesticated birds. In feral or wild fowl, a female about to lay an egg leaves her social group and normal home range and moves away to choose a nest site and build a nest or to find the site chosen on a previous day. Domestic male fowl sometimes accompany the female, and it has been suggested that both are involved in nest site selection (McBride *et al.*, 1969). An alternative suggestion, however, is that the male is interested in the nest because mating is more effective after laying than before (see Section 6.6) and in fact often occurs shortly after the female leaves the nest. In other species, the nest site may be selected and the nest built by the male alone (ratites) or by the male and female together (pigeons and quail). Nest sites are usually well defined, in places such as the foot of a slope or under a bush, secluded from disturbance and enclosed for protection from predators, although ostriches may make their nests in more open areas. If loose material is available, it may be shaped into a hollow nest bowl, but a nest may also be a simple hollow scraped in the earth or even just an area of flat ground (Duncan *et al.*, 1978; Deeming and Ar, 1999).

After the female lays a clutch of eggs, the clutch is incubated. In chickens and turkeys, the female alone incubates the eggs until hatching, but in other species (pigeons, ostriches and bobwhite) both parents participate in incubation, while in some species (e.g. rheas) only the male incubates the eggs. Incubation ('broodiness') is triggered in females by increases in the hormone prolactin and causes the cessation of further egg laying. Laying can be extended in certain wild birds by the removal of eggs, but not indefinitely: most selection for egg production has consisted of the extension of this process and hence the avoidance of broodiness (see Section 3.4).

After an egg is laid, the incubating parent sits on it only briefly before returning to normal behaviour. This remains true even when several eggs have accumulated in the nest. When the clutch is complete, however, incubation is almost continuous. When only one parent broods the eggs, that individual leaves the nest for brief periods during the day to feed, drink and defecate. When both parents brood, they each brood for a portion of the day, usually one parent during the daylight hours and the other during the night.

Only when incubation starts do the embryos begin to develop, so development is synchronous even in eggs that were laid days or weeks apart. It is important that hatching of all eggs occurs over a short period to prevent predation and because chicks are mobile soon after hatching and a brood hatched at different times would become separated. Parental behaviour, imprinting, socialization and social factors affecting hatching synchrony are considered in Chapter 5.

6.4 Sexual Development

Birds must learn the characteristics of appropriate mates for normal sexual activity. That this knowledge is not innate is demonstrated by the observation that hand-reared birds often show courtship or sexual crouching to humans. This phenomenon appears to pose a significant problem for normal reproduction in farmed ostrich (Bubier *et al.*, 1998; Soley and Groenewald, 1999). This learning, called sexual imprinting, is most likely to occur during a sensitive period prior to sexual maturity. This period is poorly defined, but in quail it occurs between hatching and 15 days of age (Gallagher, 1977), while in male chickens it is at about 10–12 weeks of age (Siegel and Siegel, 1964).

Rearing males and females separately can cause problems with breeding. Male mallards kept in captivity in all-male groups, for example, will form strong homosexual pair bonds (Schutz, 1965). Leghorn males placed in isolation or in all-male flocks during the sensitive period for sexual imprinting generally show reduced (and less successful) mating behaviour when later mixed with females (Siegel and Siegel, 1964). However, the extent of the problem depends on the rearing history of the females as well, and deficiencies in mating do decrease as the males gain sexual experience (Leonard et al., 1993a,b).

Since sexual imprinting is very important in ducks, breeding ducks are reared with the sexes together (Hearn and Gooderharn, 1988). Male and female broiler breeder chickens are sometimes reared separately for independent control of body weight, and this could result in decreased mating behaviour and fertility (Mench, 1995). Turkeys are still usually reared separately because, with artificial insemination prevalent, mating behaviour is not important. It was also thought that there was no deleterious effect of sex-separate rearing on semen production (Wood-Gush and Osborne, 1956; Siegel, 1965) but, while this may be true for semen volume, one study suggests that semen quality is poorer in males isolated from females (Jones and Leighton, 1987).

Attempts to assess the sexual potential of males very early in life have not been successful (Wood-Gush, 1963a), but tests of libido soon after sexual maturity give quite good predictions of fertility of subsequent matings (McDaniel and Craig, 1959), at least in Leghorn chickens. It may therefore be possible to select suitable males before they are used for breeding (Justice et al., 1962). However, the relationships among mating activity, semen quality and fertility can also be inconsistent, particularly in males from strains selected for rapid growth and particular conformation traits, such as broiler breeders (Wilson et al., 1979). These males could have high libido but poor fertility if their size or conformation makes it difficult for them to achieve full cloacal contact during mating (Duncan et al., 1990). In addition, even if their semen quality is good, fertility will be low if mating motivation declines due to the propensity of these males to develop skeletal problems such as osteoarthritis that restrict activity (Hocking and Duff, 1989).

In contrast to many other groups of birds, in poultry the presence of males is not necessary for sexual maturity of females. It is on this characteristic, of course, that the egg industry depends, since hens will lay large numbers of eggs in the absence of males. There is evidence, however, that egg production by turkey hens is increased by male stimulation (Jones and Leighton, 1987), so it seems that housing the sexes separately is disadvantageous for both female and male breeding potential. Similarly, the onset of lay in quail and chickens is advanced if females can at least hear male vocalizations or have visual contact with males (Guyomarc'h et al., 1981; Widowski et al., 1998).

6.5 Sexual Motivation

Individual males vary in libido, which suggests the possibilities of choosing the most effective males, as mentioned above (Justice et al., 1962), and of selecting strains for mating behaviour. Such selection has been successful experimentally, but requires careful identification of the behaviour to be selected. Thus in one study, in which overall mating frequency was used as the basis for selection, a line that mated more frequently than controls was produced. However, many of these matings were incomplete (Wood-Gush, 1960). Another study, involving more detailed observations and selection of males over 20 generations on the basis of completed matings only, met with greater success in increasing male mating behaviour (Bernon and Siegel, 1983). Unfortunately, selection for frequent mating tends to result in low semen volume per ejaculate. It may be possible to control for this while still increasing libido, but this has not been done commercially.

A male that has mated with a female may do so again if she does not move away and continues to be receptive. However, repeat matings with one female only occur after increasing intervals of time. It seems likely that such matings are advantageous in that they increase fertility, but that successive matings yield a diminishing advantage. The decline in motivation of the male is not caused by fatigue, but by habituation to the stimulus female: the male will resume active mating if another receptive female is available. This is called the 'Coolidge effect', after a well-publicized occasion on which it was explained to the American president and his wife. Habituation to particular females cannot be a problem in large breeding flocks. It is not known whether it can occur in small breeding pens or cages.

In females, there is variation between individuals and between species in how quickly they will mate again. Turkeys are more likely than hens to avoid repeat mating and may do so for some days. Since female birds can store sperm in sperm storage tubules in the oviduct, and this sperm remains viable for an extended period (Christensen and Bagley, 1989), avoidance of repeat mating is generally likely to have little effect on fertility, although turkey hens may even avoid mating after incomplete copulation or mounting by another female (Hale, 1955).

Even in large flocks, there is, of course, a limit to the number and frequency of copulations by both males and females. Each sex does, however, alter its behaviour to compensate to some extent for satiation in the other. Classic experiments by Guhl (1953) showed that cockerels are more active in courtship when hens are satiated and that hens crouch more readily to satiated cockerels.

6.6 Mating

Many farmyard flocks and small-scale operations have groups of one male with a number of females that are similar to natural harems. The nearest equivalents in large-scale production are the cages used for breeding quail and those used for chickens when selection is being carried out. Instead, breeding fowl and ducks typically are housed in large floor pens with hundreds to thousands of birds. The mating success of such a flock will be affected by the sex ratio and also by the precise housing conditions and the behaviour of the birds.

In small flocks of chickens, high-ranking males sire most of the offspring (Fig. 6.2), although the hens do mate with (and are fertilized by) more than one male (Jones and Mench, 1991). Several aspects of behaviour contribute to the mating success of high-ranking males. Females maintain closer proximity to high-ranking

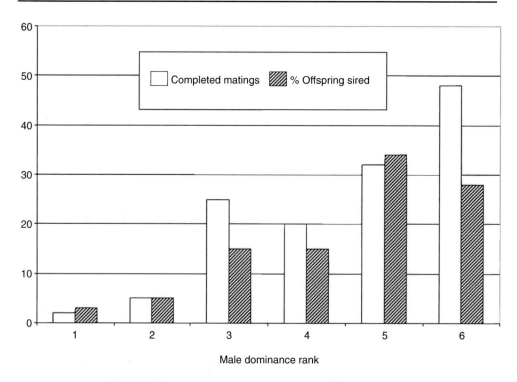

Fig. 6.2. In a small flock of chickens containing 57 hens and six roosters, the top ranking males (males with a dominance rank of 6 and 5) completed more matings, and sired more offspring, than the lower ranking males (from Jones and Mench, 1991). Paternity of the chicks was determined using DNA fingerprinting.

males (Graves *et al.*, 1985) and preferentially solicit copulations from them (Pizzari and Birkhead, 2000). It has also been suggested that hens eject the sperm of lower-ranking males immediately after copulation by those males, increasing the probability that their offspring will be sired by higher-ranking males (Pizzari and Birkhead, 2000). Males often interfere with the mating of males subordinate to themselves (Fig. 6.3). Furthermore, mating by low-ranking males can become almost completely suppressed, resulting in what has been called 'psychological castration' (Guhl *et al.*, 1945). The conditions under which this occurs are not completely understood, but probably involve small groups and crowding, which encourage a strong hierarchy among the males. High stocking density, in fact, directly restricts courtship and mating (Kratzer and Craig, 1980). However, this is not necessarily reflected in reduced fertility, at least in cages (Bhagwat and Craig, 1975).

In larger groups, rank seems to have less effect on male mating (Craig *et al.*, 1977; Kratzer and Craig, 1980). In very large groups, it used to be thought that birds would form subgroups (McBride and Foenander, 1962) that might act in a similar way to harems, but actually both males and females wander widely over most or all of the area (Appleby *et al.*, 1985). It is possible that this leads to less interference in mating than in smaller groups. However, under any conditions, because there is variation in the sexual activity of individuals, the effective sex ratio is likely to be different from the actual sex ratio. Little is known about these aspects

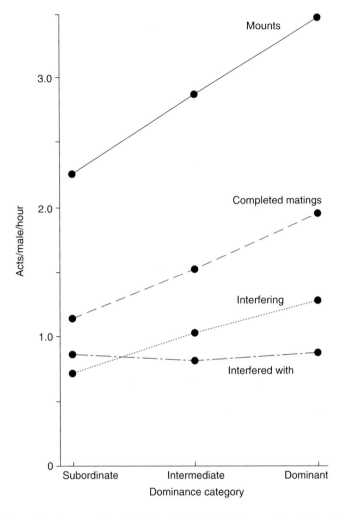

Fig. 6.3. Effect of rank on mating behaviour of White Leghorn cocks. Males frequently interfered with each other's mating (from Kratzer and Craig, 1980, with permission from Elsevier).

of reproduction, however, because there have been almost no systematic studies of mating in commercial conditions.

In addition to direct social factors, aspects of lighting can also affect mating frequency. Birds are able to see in the UV range (Section 2.4) and have UV markings that are apparently used as cues during mating behaviour. Supplementation of broiler breeders with fluorescent UV increases mating attempts by males, and females are more likely to approach and inspect males illuminated by UV light, which suggests that providing full spectrum lighting in commercial houses could increase mating activity (Jones *et al.*, 2001).

There have been few studies published of individual variation in mating frequency of male or female poultry in flocks and few even of average frequencies. A study of pens housing about 150 hens estimated that males mated about five times daily on average and that this did not increase even with a sex ratio of 24 females per male (Craig *et al.*, 1977). For females, a similar study in pens suggested they mated more than once daily (Kratzer and Craig, 1980), while a small-scale study in a commercial broiler breeder house found an average of 0.48 matings per day (Appleby and Cunningham, unpublished observations). These frequencies are higher than necessary because fertility is just as high when artificial insemination is performed less often (Craig, 1981), due to the ability of hens to store viable sperm. For example, artificially inseminating turkey hens every 2 weeks is sufficient for maximum fertility early in lay, although inseminating weekly is a more common practice (Clayton *et al.*, 1985).

Mating frequency declines with age (Fig. 6.4), but this does not result directly in lower fertility, perhaps because, as just suggested, early mating is more frequent than necessary. The decline in fertility that occurs during the laying year is, however, associated with other age-related changes in sexual behaviour (Duncan *et al.*, 1990; see next section). In broiler breeder flocks, there is a practice, called 'spiking', in which young males are added to the flock about mid-way through the laying year in an attempt to reduce this decline. It is believed that they increase the mating frequency both directly, by themselves mating, and also indirectly, by stimulating the activity of the resident, older males. No evidence has been published on the effectiveness of this practice, but in one study of such a house, the new, young males carried out a higher proportion of matings than would have been expected from their numbers (Fig. 6.5). Males may also have to be added to breeding flocks if many of them die or have to be culled, so changing the sex ratio.

Distribution of mating through the day is affected by the egg-laying cycle, because fertility is lower around the time of oviposition. Hens lay mostly in the morning. Correspondingly, they mate more frequently in the afternoon than in the morning (Upp, 1928; Craig and Bhagwat, 1974), although there is also some evidence of a peak in sexual activity after lights-on. Quails, in contrast, lay in the afternoon and mate most in the morning and evening (Ottinger *et al.*, 1982). The highest mating frequency in ostriches also occurs during the morning (Sambraus, 1994).

Although fertility and hatchability are generally high in naturally breeding flocks of farmed poultry, there are some instances in which they are poor for reasons that appear to be related to abnormalities in mating behaviour. In farmed ostriches, for example, copulation rates are low. This has been attributed to excessive aggression, sexual imprinting to humans, excessive territorial behaviour, incompatibility between males and females, poor libido or a mixture of these (Soley and Groenwald, 1999). However, there has been little research on the sexual behaviour of farmed ostrich, so there is currently little empirical support for any of these suggestions (Deeming and Bubier, 1999).

Low fertility and hatchability are also a problem in broiler breeder flocks, and both decline rapidly as the flock ages. A primary contributor to this is the tendency for these birds, and especially the males, to become obese, which affects reproductive condition (Section 6.4). However, other contributing factors appear to be related more directly to male mating behaviour. There have been recent reports of broiler breeder males directing aggression towards females, sometimes causing severe injury or even killing the females. Male to female aggression is normally rare in adult

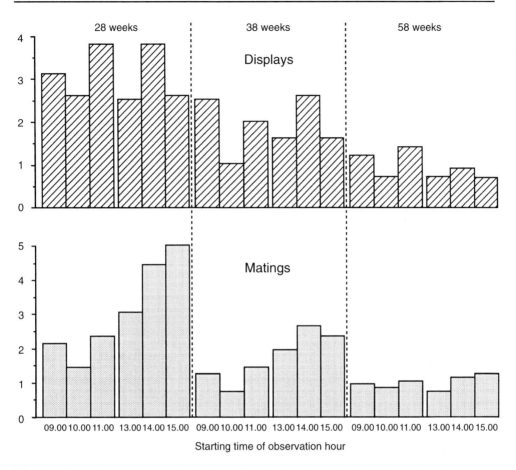

Fig. 6.4. Effect of age on mating behaviour of broiler breeder males. Displays and matings declined with age, at all times of day (from Duncan *et al.*, 1990, with permission from Elsevier).

chickens (Wood-Gush, 1956) since males and females form separate social hierarchies and males dominate females passively (Guhl, 1949). In broiler breeders, this problem results not from an increased general tendency toward aggressiveness among these males, but instead from deficiencies in male mating behaviour (Millman and Duncan, 2000a,b; Millman *et al.*, 2000). Broiler breeders are much less likely to display courtship behaviour than Leghorns, and are also more likely to chase females and force copulations on females, who often then struggle when mated. About 50% of all observed matings by these males are forced, and in the process females may sustain deep lacerations on the head and the torso, under the wings (Millman, 1999). It is unknown whether this problem is due to some inadvertent but correlated change in male mating behaviour related to selection for growth characteristics, to males resorting to forced copulations because they have difficulty completing matings normally because of their size, conformation or pain associated with leg problems, or to sex-separate rearing practices (Mench, 2002). It has been suggested that introduction of separately reared, sexually mature males into the female flock when the females are still too young to respond

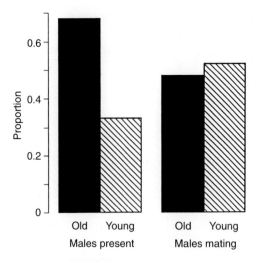

Fig. 6.5. 'Spiking' in a commercial broiler breeder flock. Three groups of young males, about 20 weeks old, were added when the original birds were 43, 46 and 54 weeks old. Observations were made when females and 'old' males were 58 weeks of age (Appleby and Cunningham, unpublished observations).

appropriately is a major cause of this problem, but there have been no published studies of this practice. The removal of toes from broiler breeder males to prevent scratching of the females can also contribute to decreased mating efficiency and fertility in these males as they age (Ouart *et al.*, 1989).

6.7 Fertility and Hatchability

The number of offspring produced by a breeding flock is the result of production of eggs suitable for incubation (i.e. clean, uncracked eggs of acceptable size, shape and shell quality), fertility (the proportion of eggs that are fertile) and hatchability (the proportion of fertile eggs that hatch). Fertility is affected by many factors (see Lake, 1969), and it will be clear from the previous section that these include behaviour and management. Hatchability is also influenced, to a lesser extent, by behaviour and management.

In breeding fowl, fertility is often 95% or higher early in the laying year, but declines sharply later, particularly after 50 weeks of age (Kirk *et al.*, 1980). This decline has been attributed to males rather than females, because fertility can be maintained by artificial insemination (Brillard and McDaniel, 1986). Duncan and colleagues (1990) therefore studied changes in the behaviour of male broiler breeders with age. They found, as mentioned in the previous section, that libido and mating declined with age, but that these changes were not accompanied by decreased fertility in all groups. In particular, fertility remained high in groups where males were slightly food restricted. It decreased, however, with males fed *ad libitum*, and the authors suggest that this was an effect of male bulk or conformation interfering with semen transfer during copulation. In other words, some matings by these birds that appeared to be complete were in fact probably not effective. Other reports also suggest that excessive weight in males reduces mating efficiency (McDaniel and Craig, 1959; Rappaport and Soller, 1966), and this problem is now usually avoided by feeding males and females separately so that male body weight can be better controlled.

It is certainly the case in turkeys that selection for body size has reduced the fertility associated with natural mating to such an extent that it is no longer viable commercially to allow turkeys to mate naturally. Artificial insemination in turkeys generally results in fertility over 85%; with proficient technique, 95% can be achieved in the first half of the laying period and 90% in the second half (Clayton *et al.*, 1985). Artificial insemination is also the rule in guinea fowl. In other species of poultry it is not generally economical, although it has been quite widely used for chickens in Israel and Japan (Cooper, 1969). Artificial insemination is also used for special breeding programmes or to perpetuate highly inbred lines with poor fertility for genetic studies (Lake, 1969).

Knowledge of behaviour can be useful for artificial insemination. For example, daily variation in insemination success closely matches the variation seen in mating frequency. In turkeys, this has been explained by the fact that contractions in the oviduct around the time of oviposition obstruct artificial insemination just as they do natural fertilization (Brillard *et al.*, 1987). This would suggest, then, that artificial insemination of turkeys and chickens should be done in the afternoon, while that of quail and ostriches should be done in the morning. Fertility by artificial insemination can also be influenced by social factors, such as the presence of males in the cage house (Ottinger and Mench, 1989).

Housing conditions also influence semen production in males. Perhaps surprisingly, both cockerels and stag turkeys produce greater volumes of semen when they are kept in cages rather than pens (Siegel and Beane, 1963; Woodard and Abplanalp, 1967). However, this may actually be an effect of group size, with male–male mounting in groups reducing the remaining semen volume compared with isolated males (Ottinger and Mench, 1989). Crowding and large group sizes should be avoided, partly because stress can depress reproductive ability (Ottinger and Mench, 1989). Separation of the sexes also affects semen quality, as already mentioned: quality is poorer in non-mating turkey males than in those allowed to mate naturally (Jones and Leighton, 1987).

Several of the factors that affect fertility also affect hatchability, including certain aspects of behaviour and management. For example, hatchability varies for eggs laid at different times of day. It also declines with time after insemination, especially towards the end of the fertile period (Landauer, 1967), due to the decreasing ability of the sperm to inseminate the egg and an increase in early embryonic mortality.

6.8 Control of Egg Laying

Much of the behaviour associated with egg laying is under genetic control. This may contribute to the very consistent, even rigid, way in which this behaviour is expressed, although environmental factors, including the housing system, can modify details of the behaviour.

Genetic effects are also evident from differences between strains in the details of pre-laying behaviour (see Section 6.11) and from the reduction in broodiness caused by selection for egg production (Section 3.4). There may have been incidental selection in the past for or against certain aspects of laying behaviour. For example, in farmyard flocks, those birds nesting in inaccessible sites were more likely to rear a brood than those laying in nest boxes, in which the eggs were taken away from them. Conversely, in more modern systems, where breeding birds are enclosed in houses with nest boxes, there is indirect selection against floor laying. Breeders use trap nests to identify which birds lay which eggs: these nests have a lever over the entrance that triggers a door to close when a bird enters. The door remains shut until it is opened manually and the egg labelled. Eggs laid on the floor cannot be attributed to the birds that laid them and these birds are therefore likely to be culled as poor layers. Even without trap nesting, floor eggs are more likely to be broken or dirty than nest eggs and therefore are rejected for incubation. Such indirect selection can have had only a minor effect, however, because the causes of floor laying are complex (see Section 6.10).

While the actual genetic mechanisms that control nesting are not fully understood, a considerable amount is known about physiological control. It might be expected that pre-laying behaviour would be triggered by the presence of an egg in the shell gland, ready to be laid. In fact, it is triggered by ovulation approximately 24 h earlier (Wood-Gush and Gilbert, 1964) and the release of oestrogen and progesterone from the follicle after ovulation (Wood-Gush and Gilbert, 1973). These hormones act on the central nervous system and cause nesting behaviour to be initiated after a suitable time interval. Meanwhile, the egg is developing quite independently of this process (Section 2.10) and it is ready to be laid by the time nesting behaviour has begun. Pre-laying behaviour and oviposition are therefore usually synchronized appropriately (Fig. 6.6).

One consequence of this mechanism, however, is that once ovulation has occurred, pre-laying behaviour will proceed even if something goes wrong with normal development of the egg: it will start at the expected time, but without an egg to be laid. The most common cause is 'internal laying', where the ovum is not picked up by the oviduct and enters the peritoneal cavity where it is resorbed internally; alternatively, sometimes an egg is laid prematurely without a hard shell (Wood-Gush, 1963b). These problems occur at quite a high frequency in all systems, but are largely unrecognized because behaviour is not usually recorded. One study in which nesting behaviour was recorded using trap nests, though, suggested that up to 12% of potential eggs were being lost because of internal laying (Wood-Gush and Gilbert, 1970). It would certainly be possible to investigate this problem further in commercial strains, and perhaps select against it, by studying the frequency of nesting without laying.

Once pre-laying behaviour has been triggered, birds have very strong motivation to find a suitable place for laying (Duncan and Kite, 1987). What constitutes a suitable place and how this is reflected in the behaviour of the birds are discussed in Section 6.10.

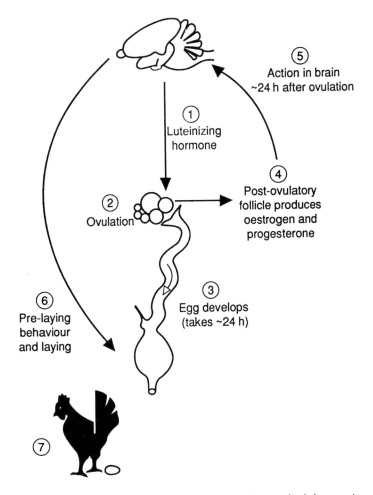

Fig. 6.6. Control of pre-laying behaviour. Ovulation in the single, left ovary is under endocrine control. While the egg develops in the oviduct, there is then hormonal feedback in the brain, which triggers the behaviour at an appropriate time. See also Section 1.14.

6.9 Timing of Egg Laying

The fact that ovulation in chickens occurs around dawn and oviposition about 24 h later means that the timing of egg laying is strongly influenced by the ambient light in open systems and by the lighting regime in closed houses. Since it is triggered by ovulation, pre-laying behaviour therefore can only occur during a certain period (Fig. 6.6). If oviposition is delayed beyond this period, it is not accompanied by pre-laying behaviour and the egg is laid in the course of other activity, often with hardly even a change of posture. The most common cause of such a delay is social interference between birds. This may occur in floor systems when the nest boxes are all occupied, particularly if high-ranking birds either prevent low-ranking birds from entering one or more boxes (Perry *et al.*, 1971) or otherwise direct aggression towards them. Aggressive interactions directed towards subordinates increase during the period immediately prior to laying, and subordinate hens receiving aggression spend less time sitting on the nest and are more likely to be displaced from nests (Freire *et al.*, 1998; Lundberg and Keeling, 1999). To ensure that there is sufficient nest space for all hens to nest undisturbed and thus prevent floor laying, it is thus important to provide sufficient nest boxes (Appleby, 1984). A ratio of one nest for every four or five birds is usually recommended, and a survey of broiler breeder houses found fewer floor eggs with this number than with higher ratios (Brockle-hurst, 1975). In practice, ratios of from 1:6 to 1:8 are common. Delay of oviposition may also be caused by human disturbance during nesting.

It is also possible that some management practices contribute to delay of egg laying, particularly feeding birds during peak laying time. When limited food is provided on a schedule, feeding behaviour of hungry birds sometimes suppresses their pre-laying behaviour and restricted-fed birds are sometimes seen to lay while feeding. In one study of broiler breeders, however, varying feeding time had no effect on the proportion of eggs laid on the floor (Hearn, 1981).

6.10 Nest Site Selection

All housing systems for laying poultry, except conventional battery cages for hens and quail, involve the collection of eggs from nest boxes. The behaviour associated with birds choosing where to lay is therefore critical in such systems. It also has important effects in cages, since eggs laid at the rear of the cage are more likely to be cracked as they roll forward than those laid at the front. In cages with perches, laying from the perch is another cause of cracked eggs (Duncan *et al.*, 1992).

Failure to use nest boxes can be a major economic problem in most or all non-cage systems and potentially in some modified cage designs (Sherwin and Nicol, 1993). In severe cases, 50% or more of eggs are laid on the floor. These eggs are labour intensive to collect, and attempts to prevent floor laying also involve much work. Floor eggs are often broken, which encourages egg eating, or dirty, which reduces value or hatchability (Hodgetts, 1981). Floor laying is variable, not just between systems but within systems, even from flock to flock or from pen to pen, and this variation has in the past seemed intractably unpredictable. However, there is increasing understanding of the factors that affect nest site selection and floor laying. These include rearing conditions, housing conditions, nest box design and management and human intervention (Appleby, 1984; Sherwin and Nicol, 1993).

The conditions in which birds are reared affect later nest site selection in two ways. First, they affect development of mobility, which has important effects in many poultry on the birds' ability to gain access to nest boxes. In particular, some strains of hens, including at least medium hybrid layers and broiler breeders, learn more readily to jump or flap from the ground to higher levels if they have the opportunity to do so when they are young (Fig. 6.7; Faure and Jones, 1982; Appleby and Duncan, 1989). Most nest boxes for adult hens are raised above ground level, so hens reared with no experience of perching, either in cages or on litter, lay many floor eggs as adults (Craig, 1980; Appleby *et al.*, 1986). In some rearing houses, the problem is exacerbated further by the use of electric wires to prevent birds sitting on feeders, effectively training them not to perch. Avoidance of such mistakes and provision of perches during rearing can greatly reduce later incidence of floor laying (Fig. 6.7; Appleby *et al.*, 1983, 1988a; Brake, 1987).

Secondly, rearing conditions may influence later nest site preferences. Most such effects are likely to be subtle, e.g. experience of dark or bright conditions affects choice of dark or bright nest boxes (Wood-Gush and Murphy, 1970; Appleby *et al.*, 1984b). However, it has also been suggested that allowing immature birds to investigate boxes increases their readiness to use them later (Rietveld-Piepers *et al.*, 1985). Certainly floor laying is worse if birds are kept in rearing houses until after maturity and establish laying patterns in the absence of boxes (Dorminey, 1974; Sherwin and Nicol, 1993).

Conditions in adult laying houses also affect mobility of birds and thus their use of nest boxes. In deep litter houses for laying hens, if drinkers or roosting areas are raised above ground level, this encourages birds to perch (Appleby, 1984) and increases use of raised nests (Fig. 6.8; Maguire, 1986). On the other hand, floor laying is a major problem in some aviaries (Hill, 1986) and percheries (Anonymous, 1983), even though hens in these systems must be able to perch. It is possible that other factors are affecting the accessibility of nests in these houses. Other aspects of housing that have been suggested to influence floor laying include lighting, floor material and temperature, but evidence for these effects is equivocal (Appleby, 1984).

The fact that mobility of at least some types of poultry is restricted means that accessibility of nest boxes is one of their most important features. Ground-level boxes may be advantageous, although some birds may even have difficulty using

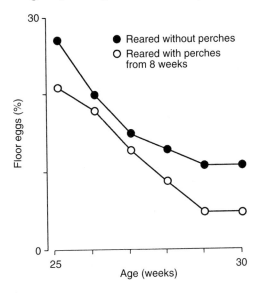

Fig. 6.7. Effect of providing perches during rearing on floor laying in commercial flocks of broiler grandparents. In the experimental flock, perches were provided later than ideal, 8 weeks after hatching, but this still considerably reduced floor laying with raised nest boxes (from Appleby *et al.*, 1988a, with permission from Taylor & Francis Ltd).

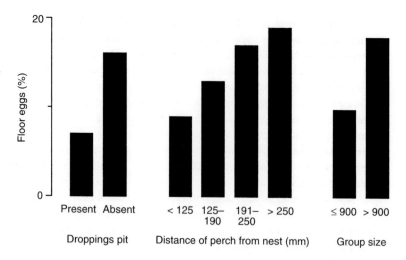

Fig. 6.8. Effect of housing conditions on floor laying: average results from a survey of 53 commercial pens of broiler breeders (after Brocklehurst, 1975, with permission from Taylor & Francis Ltd).

these if they have to step up into them (Appleby *et al.*, 1988a). With raised boxes for hens, easy access from an alighting rail or similar arrangement is essential (Fig. 6.8; Appleby, 1984). In fact, the high proportion of hens that do use raised boxes is quite surprising, since hens and related birds nest on the ground in the wild. A likely explanation is that the main characteristic such birds use to select a nest site is enclosure, or protection. Nest boxes are more enclosed than any natural sites and so are usually chosen in preference to positions on the floor (Appleby and McRae, 1986). If there are enclosed sites on the floor, however, in corners or under nest boxes, these may be as attractive as the boxes themselves. Houses should be arranged to avoid providing such sites. The requirement for enclosure is not stringent, and most designs of nest boxes are sufficiently enclosed. Indeed, the fact that many different designs of nests have been successful (Fig. 6.9) suggests that the important features are quite simple (Smith and Dun, 1983).

Fig. 6.9. Some common types of nest box. From left to right: wooden boxes with wood shavings as litter; metal boxes with plastic rollaways; wooden boxes with 'AstroTurf' rollaways; and metal 'autonests' with buckwheat litter on a conveyor (from Appleby *et al.*, 1988c, with permission from Taylor & Francis Ltd).

Features often thought to be important such as darkness and seclusion may affect which nest boxes birds choose (Fig. 6.10), but are unlikely to affect the choice between nest boxes and the floor. Similarly, although birds prefer some nesting materials to others (Huber *et al.*, 1985) and prefer nests containing eggs to others (Kite *et al.*, 1980), there is no evidence that these factors influence floor laying. Only if there is no nesting material at all does floor laying seem to be worse. This is particularly important with regard to automatic egg collection systems for hens. Rollaway nests, which allow eggs to roll into a collection channel, are sometimes very successful, but less reliably so than littered nests. This is probably partly because of the lack of nesting material, despite the use in some designs of artificial grass, or AstroTurf, as a yielding substrate for hens to nest on. The sloping base and rolling away of eggs which are integral to these nests are probably also aversive: blocking rollaways during early lay increases their use (Appleby, 1990). Nevertheless, nests with egg collection systems incorporating litter are potentially more appropriate for floor-housed poultry (Appleby *et al.*, 1988c), since hens prefer loose nesting material if it is available. Prior to egg laying, they will choose to build a nest in such material rather than using a moulded plastic nest or a nest that they themselves previously formed (Duncan and Kite, 1989). However, the interpretation of this preference is difficult. Wild or feral birds do not nest in deep, loose material, and it seems likely that, as with enclosure, hens are reacting to stimuli that are stronger than they would encounter in natural conditions. Nevertheless, provision of loose nesting material for hens is often recommended on welfare grounds, since hens seem to be motivated to construct a nest shortly prior to egg laying (Hughes *et al.*, 1989).

Fig. 6.10. Traditional ideas about poultry behaviour are not always right. In this experiment, some hens chose the dark nest box, as expected, but others chose the light one (from Appleby *et al.*, 1984b, with permission from Elsevier).

While all these aspects of management have indirect effects on nest site selection, humans may also influence pre-laying behaviour directly, by training the birds. Most laying farms for hens carry out some training when the birds are first housed, in an attempt to prevent floor laying becoming established. Some methods could be called negative, such as disturbance of birds sitting on the floor and destruction of floor nests; it seems unlikely that these measures are effective. Positive methods, such as placing birds into nests, are more likely to work, especially if the birds can be confined there for a while: this is possible in trap nests and in some other designs with hinged perches. In one trial of this method, floor eggs laid by hens confined briefly in nests subsequently declined to 1%, compared with 24% in control pens (Craig, 1980). Even confinement for a period as short as 30 min greatly reduced floor laying in experimental flocks (Fig. 6.11). However, it is probably rare for training in commercial flocks to be carried out systematically. Reduced attention to individual birds in large flocks may explain the finding that floor laying is worse on average in pens of more than 900 birds than in smaller pens (Fig. 6.8).

Nest site selection is thus a complex process, but careful attention to all stages of the life history of the hen should allow it to be controlled satisfactorily. Once problems such as floor laying have become established, however, the conservatism of nesting birds often makes such problems difficult to cure. For hens housed in modified cages, the incidence of floor laying can be decreased by transferring the

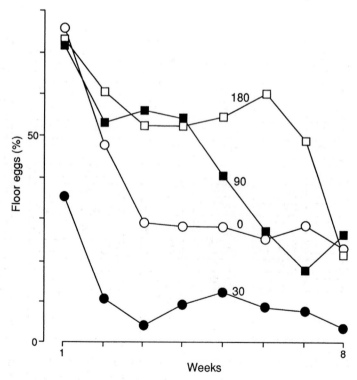

Fig. 6.11. Effect of confining hens in nest boxes on subsequent floor laying. In this Australian experiment, groups of broiler breeder hens were shut into nests for 0, 30, 90 or 180 min (Maguire, 1986).

hens to cages at an early age so that they have experience of the nest box prior to onset of lay (Sherwin and Nicol, 1993).

6.11 Pre-laying Behaviour

In housing systems with littered nest boxes, complete pre-laying behaviour similar to that in natural conditions is shown, with a searching phase, choice of a nest site and creation of a nest hollow (Wood-Gush, 1971). However, behavioural problems may still occur in such systems, depending on pen size and the number of nest boxes. If nests are limited, aggressive interactions are common (Meijsser and Hughes, 1989) and floor laying may occur. In contrast, in large pens with many similar nest boxes, hens have difficulty choosing among them and sometimes show pacing behaviour, which suggests that this difficulty is frustrating and deleterious to welfare (Appleby *et al.*, 1986). They also lay in different boxes on successive days (Fig. 6.12).

Difficulty in choosing between nest boxes probably also accounts for another feature of pre-laying behaviour that occurs in pens, namely gregariousness (Fig. 6.13). Domestic hens and female turkeys will often enter occupied nests even if there are others free (Appleby *et al.*, 1984a), and for this reason colony nest boxes rather than individual nest boxes are now often used in commercial laying hen houses. In

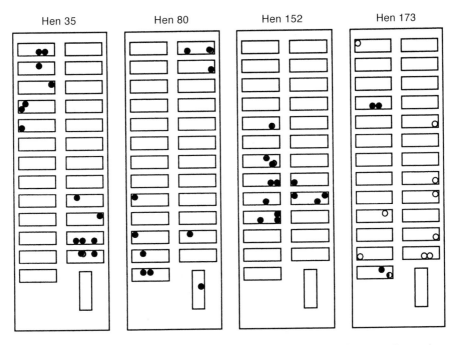

Fig. 6.12. Nest choice by commercial broiler breeders. Nesting behaviour of tagged hens was recorded on a number of days in a flock of nearly 4000 birds. Nest boxes were in blocks of 24, with two tiers of six on each side; they are shown here exaggerated in size. Open symbols = upper tier; filled symbols = lower tier (from Appleby *et al.*, 1986, with permission from Taylor & Francis Ltd).

Cages

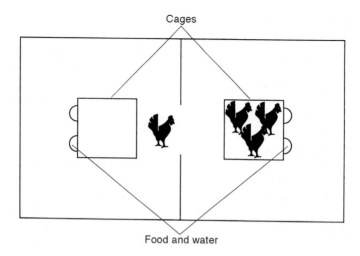

Food and water

Fig. 6.13. Gregarious nesting behaviour. In this experiment, hens were tested singly, with the choice of laying near flockmates or in isolation. The majority laid in the pen containing their flockmates (from Appleby *et al.*, 1984a, with permission from Taylor & Francis Ltd).

extreme cases, however, gregarious nesting can lead to breakage of eggs (which encourages egg eating) and even to birds being suffocated. It seems to arise very early in lay, in birds confronted with almost identical nest boxes. Occupied nests are the only ones that contrast with others and are investigated preferentially (Appleby and McRae, 1986). All these effects will be reduced in smaller pens, or in more heterogeneous houses that allow birds to localize their activities. Gregariousness is also avoided in turkeys by the use of semi-trap nests, with doors that prevent access by additional birds. These have not been used for hens, probably on grounds of cost.

   In rollaway nest boxes, birds will usually settle as normal, but the behaviour associated with nest building is reduced in the absence of litter. Some rollaway nest boxes, however, stimulate nest site selection but not settled nesting: some birds will enter such boxes many times in a frantic manner before laying (Appleby, 1990).

   Such frantic pre-laying behaviour is, however, more characteristic of laying hens in conventional battery cages, where the searching phase is extended, in the form of restless pacing around the cage. Subsequent phases, however, differ between strains of hens (Wood-Gush, 1972). In most light hybrid strains, settled nesting is very brief or completely absent, and hens continue to pace, often in a stereotyped way, until shortly before or even after oviposition. Frustration of nesting is a severe behavioural problem for hens in cages. The extent of pre-laying pacing is under genetic control: its incidence can be reduced by appropriate selection (Mills *et al.*, 1985a), and differences in the duration of pre-laying behaviour are seen in different commercial light hybrid lines (Heil *et al.*, 1990). However, it is not known whether the lines that show decreased pacing are actually under less stress, or whether stress is simply not expressed in the same way. Medium hybrids, in contrast, usually sit before laying and often show 'vacuum' nesting behaviour: they go through the motions of making a nest hollow even though there is no litter. It has sometimes been suggested that this shows they are highly motivated to nest and that they are likely to be frustrated by absence of an appropriate substrate. However, this is not supported by records of heart rates, which suggest that the birds are calm during vacuum nesting (Mills *et al.*, 1985b). One other possibility is that they have little need of such a substrate (Appleby, 1990). At any rate, the calmer behaviour of medium hybrids compared with light hybrids in the pre-laying period is generally interpreted as better adaptation to the cage environment.

Frustration of nesting in cages can be avoided by the addition of nest boxes to modified designs (Robertson *et al.*, 1989; Appleby and Hughes, 1990; Appleby *et al.*, 1993), but the problems of automatic egg collection in such designs have not yet been wholly solved (Abrahamsson and Tauson, 1997).

Different strains of hens also show variation in other aspects of egg laying. In particular, floor laying is more common in medium hybrids than in light hybrids and in broiler breeders than in layer breeders. This is a reflection of the difference in mobility between these categories of bird and requires appropriate management (Section 3.5).

6.12 Oviposition

Behaviour during egg laying itself may be an important cause of damage to eggs, particularly in cages. Individual birds vary in their laying stance, some habitually standing to lay, which tends to result in cracked eggs (Carter, 1971). In cages with perches, hens may lay from the perches, perhaps because these are more level than the sloping floor, or because they provide more support, and this increases the problem (Duncan *et al.*, 1992). Although the percentage of floor eggs seems to be decreasing in modified caging systems as nest box designs are improved, the high proportion of cracked eggs is still a serious problem in these systems, although the percentage of cracked eggs can be decreased by providing egg savers or nest curtains (Wall and Tauson, 2002). The problem is generally absent in systems where eggs are laid in nest boxes containing litter or some other soft surface. An exception to this is in severe cases of gregarious laying when eggs are laid on top of other eggs.

In nest boxes, laying position can be a critical factor in the initiation of cannibalism (Section 5.10). Vent pecking is more likely when birds face inwards to lay; this seems to be particularly common in rollaway nests that slope forwards, but it is not clear why birds should prefer to face up the slope rather than down.

If oviposition is delayed, retention of the egg in the shell gland often causes deposition of extra calcium on the surface. This gives a 'dusted' appearance, which is harmless to consumers but may nevertheless reduce the price paid by some buyers. It probably also reduces gaseous exchange through the shell and hence hatchability of fertile eggs. It sometimes occurs naturally, particularly early in lay, but is more frequent after disturbance during the pre-laying period. Disturbance to birds at an earlier stage also results in abnormal eggs, but in this case they are usually mis-shapen, presumably because of contractions in the shell gland before the shell has hardened (Figs 6.14 and 6.15).

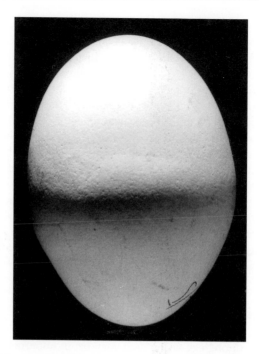

Fig. 6.14. One of the possible effects of disturbance to birds while the shell of the egg is still soft: an equatorial bulge, caused by contraction of the shell gland.

6.13 Post-laying Behaviour

In cages or rollaway nest boxes, birds have little opportunity to sit on their eggs, although they sometimes do so for a while if eggs fail to roll away immediately. They also sometimes continue to pay attention to eggs even after they have rolled away. In littered boxes, in contrast, birds sit on eggs for a variable period, up to about half an hour. There is some indication that this results in shorter shelf life for table eggs, by preventing them from cooling down so quickly. Sitting is also extended in rollaway nests if false 'decoy' eggs are fixed in them to encourage their use. In either of these circumstances, nests are occupied for longer, and so a sufficient number of nest boxes must be provided to account for this.

Allowing birds to sit on eggs also increases the likelihood of broodiness, even in highly selected laying strains. Appropriately, then, the usual treatment for broodiness in laying birds from a floor system, especially in turkeys, is to shut them in a cage or 'broody coop' on wire or slats. In most cases, they quickly resume laying. If broodiness is required, however, for example for incubation in small, farmyard flocks, it can be induced by introducing chicks to the females.

One other problem which sometimes arises with littered nest boxes is egg eating. It is always initiated by accidental breakage of eggs, for example after floor laying or overcrowding of nest boxes, but once birds have experience of eating broken eggs they may learn to break more themselves. The problem is rarer in systems where eggs roll away, but not unknown: if eggs are cracked in battery cages, for example, hens can learn to peck them before they roll out of reach. As with

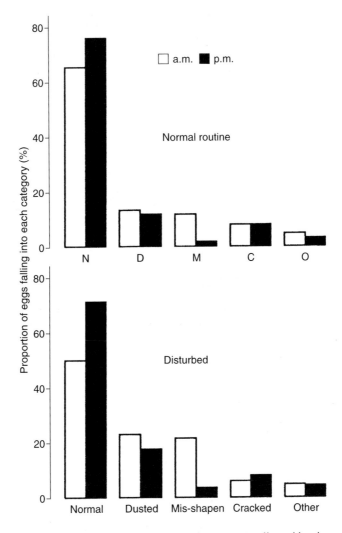

Fig. 6.15. The effect of disturbance to birds on their eggs is affected by the stage of egg development. The incidence of shell abnormalities from hens housed on deep litter was recorded on days following a normal routine (upper diagram) and following a major disturbance (lower diagram). Most eggs collected in the morning (open columns) would have been in the shell gland during the disturbance and a high proportion were mis-shapen. Eggs collected in the afternoon (filled columns) would have been ovulated later (Hughes *et al.*, 1986, with permission from Taylor & Francis Ltd).

other behavioural problems, this is difficult to cure once it has become established, so it is important to prevent it by good management. Reduction of light levels, however, may help to reduce the problem. Conversely, bright light seems to be a contributing factor, perhaps through a general effect on activity.

Table 6.1. Several studies in the 1980s compared egg numbers in cages with those in other systems, all using ISA Brown hens. Not all results were tested statistically, and different studies are not directly comparable.

Reference	System	Ages (weeks)	Stocking density (birds/m²)	Year 1	Year 2	Year 3
Hughes and Dun (1986)	Free range	20–68	0.1 (3 in house)	245		
		20–72			283	287
	Cages	20–68	21	251		
		20–72			280	284
Gibson *et al.* (1988)	Strawyard	20–72	3	261		
			4	247	288	291
			5		283	
			6			285
Appleby *et al.* (1988b)	Deep litter	20–64	3	224		
			6	228		
			7		208	
			8		224	
			10		235	225
			11			232
	Cages	20–64	13	242	234	253
			18	236	230	248
McLean *et al.* (1986)	Perchery	20–44	13	133		
	Cages	20–44	18	137		

6.14 Egg Production

Although food intake and food conversion efficiency vary between systems (Section 14.3), no clear differences in egg production have been found. For laying hen production systems, however, a difference that does exist is loss of eggs in more extensive systems, from causes such as floor laying. As a result, there has been a slight tendency for more eggs to be collected in cages than in other systems. This has not been consistent (Table 6.1) and should be amenable to management.

Within systems, it is common for egg production to decline with increased stocking density and group size. For laying hens, this has been most clearly demonstrated in cages (Hughes, 1975b; Adams and Craig, 1985); it has also been recorded in a strawyard, but not in deep litter (Table 6.1). Depression of egg production has sometimes been interpreted as an indicator of poor welfare; this is controversial (Appleby and Hughes, 1991), but in caged hens decreased egg production is correlated with more reliable indices of welfare, such as mortality (Adams and Craig, 1985).

Egg production is unique in animal production systems. The product, whether used for human consumption or for breeding, is conveniently packaged and has an integral delivery mechanism in the form of 'pre-programmed' egg-laying behaviour.

This behaviour is also unusual in the extent to which we understand it, so it is ironic that a desire to control this behaviour was a major impetus for the development of battery cages. Such understanding can now be put to good use in management of other systems.

7 Perceptions of Welfare

7.1 Summary

- The interests of animals and of humans sometimes coincide and sometimes conflict. Nevertheless, most people consider that animal welfare is a valid concept and that animals have some intrinsic rights.
- Concerns about animal welfare increased following the switch from floor systems to cages that gained pace in the 1960s. Polls in the UK over many years have indicated support for less intensive systems, among farmers as well as the public. Pressure to ban caging has, nevertheless, mostly come from city dwellers, and there are also substantial differences between cultures and countries.
- Philosophers have outlined ethical approaches to animal welfare. A utilitarian approach weighs up the consequences of actions to determine which yields the greatest good for the greatest number. This contrasts with the approach in which actions are themselves considered acceptable or unacceptable, regardless of consequences. Here there is an emphasis on human duties and on animal rights. Most people have views that include elements of both these approaches.
- When adopting an animal-centred approach, it is important to distinguish between factors that affect welfare, such as provision of food and water, and the effect of those factors on the animal itself, such as comfort and freedom of movement.
- Duncan and Fraser have argued that there are three approaches to welfare, emphasizing feelings such as pleasure and suffering (mind), avoidance of injury and disease (body), and ability to express natural behaviour (nature).
- The idea that animals should be allowed to express natural behaviour in a natural environment is difficult to translate into specific recommendations. However, to respect an animal's nature means taking a broad approach, identifying factors important to it and meeting its needs as far as possible.

7.2 Human and Animal Interests

In most of our interactions with animals, there is at least some overlap between our interests and theirs; for instance, it is to the benefit of both a farmer and his/her stock if the latter are healthy. Many of the practices involved in poultry keeping are based on this mutual benefit, and recognition of this can be thought of as enlightened self-interest on the part of the farmer. There are still many ways in which this approach can be taken further. For example, in many hot countries, there is considerable room for improvement in cooling of poultry houses, which will reduce heat stress in birds and increase their profitability (Sainsbury, 2000).

Unfortunately, the overlap between human and animal interests is not complete. Some farming practices are not beneficial to the animal, such as force feeding of ducks for the production of fatty livers. Also, the farmer in general is concerned with animals as a group rather than as individuals. So even if a particular practice leads to decreased performance of an individual, it may increase economic performance overall: thus increasing stocking density of laying hens in cages often reduces the laying rate and food conversion efficiency of individual birds but increases economic output of the house. These types of conflict have given rise to the need to consider animal welfare as a separate issue from that of the human benefits of animal use.

There are several different approaches to understanding animal welfare concerns. Contradictory though this sounds, one school of thought holds that animal welfare is human centred, i.e. that the concept has no meaning except in the context of our interactions with animals. An extreme form of this view was that of the philosopher Kant (1786) who held that the reason cruelty to animals is wrong is that it makes the perpetrators more likely to be cruel to other people. That idea is now outdated, but some current writers suggest that we can never know definitively what matters to animals, and that as we are in control of our interactions with animals, it is only our view of welfare that actually has any relevance (Carruthers, 1992; Kennedy, 1992). Similarly, some economists take the practical approach that humans are making the decisions and that it is therefore only useful to consider animal welfare in so far as it matters to humans (McInerney, 1994; Bennett, 1997). If people are willing to expend time, effort or money on improving conditions for animals, a value can be put on that willingness and it can be balanced against other values such as short-term profit (Fig. 7.1). If not, any consideration of welfare is academic. This human-centred approach is also shown by some people who consider that welfare refers only to our obligations to animals. Andrew Fraser (1992) uses welfare in this sense and the alternative term well-being to refer to factors intrinsic to the animal. This distinction is not common. Welfare and well-being are treated in this volume as synonyms.

The human-centred approach to animal welfare may be practical in some circumstances, but it is not widespread. Most people feel that welfare is rooted in the animals themselves, e.g. that wild animals have 'welfare' even if they do not interact with humans. Thus it is possible to compare the welfare of a wild pheasant with one in captivity. It is true that we have to make human judgements about what is important for animals (see Section 7.7), but this involves understanding animals and their welfare, by examining to at least some extent their point of view. Most people,

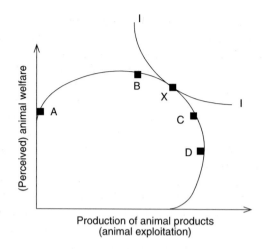

Fig. 7.1. Putting a value on animal welfare. Imagine that humans are starting to exploit animals at point A. This model assumes that up to point B, animals and humans derive mutual benefit from their association. B marks maximum welfare for animals with some benefits for humans. However, maximum output of animal products for human benefit would be achieved at point D, at a cost to animal welfare; exploitation beyond this point would reduce production. The decision for society is where on the curve from B to D should we be? Arguably, society is operating at C. This implicitly gives a relative value to animal welfare, because it implies that society forgoes a unit of animal product to gain a unit of animal welfare. However, we may feel that point C does not accurately reflect people's concerns for welfare and if we really knew society's preferences – shown by function I – then we should rather be at point X (i.e. achieving a lower level of production but with higher welfare) (Bennett, 1997).

then, see animal welfare as animal centred, reflecting the idea that animals have interests, or intrinsic value, or rights. This volume takes this animal-centred approach and examines how the interests of our domestic poultry are affected by the ways in which we keep them.

7.3 Public Attitudes

The welfare of domestic poultry has been a controversial issue in developed countries at least since the early 1960s, when systems for laying hens such as deep litter and free range were being phased out and replaced by cages. Public awareness of intensive husbandry methods was increased by the publication in the UK of *Animal Machines* (Harrison, 1964) and the Brambell Report (Her Majesty's Stationery Office, 1965), and has been maintained by the activities of animal protection societies and concerned professionals including scientists. Politicians in several countries say that the single issue about which members of the public write to them most is animal welfare.

These general impressions about public attitudes are supported by polls. Over the same period, polls have suggested that a majority of people are unhappy with intensive husbandry systems. In 1968, about 90% of people in the UK felt that poultry should have enough room to spread their wings and 79% considered that all livestock, including poultry, should have access to the open for some part of the day during fair weather (Social Surveys, 1968). In the USA, even though most polls show that the public is supportive of farmers and ranchers, there also appears to be growing concern about intensive housing of animals (Swanson and Mench, 2000). Recent national polls (Caravan Opinion in 1995 and Zogby America in 2000: see Swanson and Mench, 2000) indicate that approximately 90% of those surveyed disapproved of confinement systems for veal calves and pigs, and of keeping hens in cages that do not allow them enough space to stretch their wings. It is important to remember, however, that the results of such surveys are affected by the precise questions asked. In 1983, National Opinion Polls found that 90% were in favour of legislation to improve conditions for intensively housed animals. In the same year, another survey (Market and Opinion Research International, 1983) asked a different question – whether intensive housing should be banned. A smaller (but still considerable) proportion of 47% of people answered 'Yes' to this question, while the same proportion felt that intensive housing is unavoidable.

It is often implied that farmers do not share this general concern for animals, but this is unfair. Most people who keep animals do so because they like animals, and most show at least the enlightened self-interest discussed above. The 1968 UK survey found that a substantial proportion of farmers polled held opinions similar to those of the general public. In a more recent UK pilot study into the attitudes of people towards different housing systems for laying hens (Rogers *et al.*, 1989), three groups of people, categorized as agriculturalists, welfarists or general public, completed a questionnaire after watching a video of the salient features of six housing systems. There were striking similarities between the groups, but also some differences. All apparently preferred less intensive systems, with free range given the highest rating and cages the lowest (Fig. 7.2). However, the groups listed different factors as important when deciding how to house hens. Economics was regarded as the most important factor by the agriculturalists, whereas the welfarists and the general public placed the behaviour of the birds first. Agriculturalists tended to assess all the systems as better than welfarists did, with less spread between the best and the worst. The reactions of the general public were intermediate. Similarly, interviews in The Netherlands showed that farmers who housed their livestock in intensive systems felt that they treated their animals well because the animals were healthy. However, consumers felt that welfare was poor because the animals lacked freedom to move and to fulfil their natural desires (Te Velde *et al.*, 2002).

Because these were small-scale studies, the results should be viewed with caution, but they do suggest that some underlying attitudes, such as a marked preference for less intensive methods of poultry husbandry, are firmly rooted in the population.

It remains true that producers are more affected by the economics of poultry production than are other members of the population, so the balance they strike between human and animal interests is likely to be different. However, even the attitudes of the public are constrained by other factors such as food prices. This will be discussed further in Chapter 14.

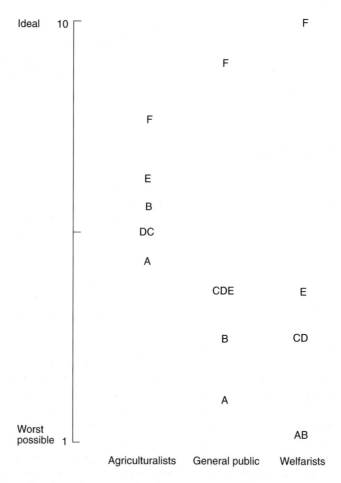

Fig. 7.2. Overall ratings given to poultry housing systems by three categories of people after watching a video in which the six systems were shown. A, battery cages; B, modified cages; C, perchery; D, deep litter; E, covered strawyard; F, free range (Rogers *et al.*, 1989).

In the Market and Opinion Research International poll (1983), those with the least experience of agriculture favoured a ban on intensive housing more than those who had close contact with farming. This fits a general pattern in which concern for animal protection has largely come from city dwellers rather than from those directly involved with farm animals. Correspondingly, this concern historically has been strongest in northern and western Europe, in countries such as the UK and The Netherlands, which are more industrialized than others (Chapter 13). Concern is also strong in areas such as North America (particularly Canada) and Australasia, presumably because emigration from Europe led to cultural similarities.

Other cultural differences are more difficult to explain. In some countries such as Japan, there is less concern for animal welfare than in, say, Europe, but more about the killing of animals. So while many families have their own rice paddies, few rear animals such as ducks that they themselves would have to kill. Sick poultry are sometimes allowed to die slowly rather than being culled, because people are reluctant actually to kill them. Some of this is religious, part of the attitudes to life and death of Buddhism and Shintoism, but that is not a full explanation.

There are also clearly changes in attitudes towards animals within societies over time. The next section will consider the ethical basis of attitudes and of changes in attitude.

7.4 Ethical Approaches

One reason why people vary in their ethical judgements is that there are different approaches to ethics. Some moral philosophers argue or imply that people should adopt one approach in full, so as to be as consistent as possible in their ethical decisions. However, others reject the idea that a single approach is appropriate. In everyday life, people tend to have hybrid views, accepting elements from two or more approaches.

One major approach concentrates mostly on the consequences of our actions. The best known form of this is utilitarianism: the idea that we should act so as to produce the greatest good (utility) for the greatest number of individuals. When Jeremy Bentham (1789) developed utilitarianism, he only considered utility for humans, but in recent years the idea has been extended to animals, particularly by Peter Singer (1975). Singer suggested that when we reckon up the good and bad consequences of our actions, consequences for animals must be given equal weight to those for humans unless there are impartial grounds for doing otherwise.

Few people are straightforwardly utilitarian. For example, most feel that the pleasure of thousands of spectators did not justify the death of one gladiator in Roman times, and that enjoyment of spectators does not justify the injuries caused by cockfighting today either. There are other problems in applying utilitarianism. Reckoning up good and bad consequences is often difficult or impossible, particularly because those consequences are frequently unpredictable. For example, is beak trimming to prevent cannibalism (Chapter 5) justifiable, if the chance of cannibalism is unknown? Are the welfare problems caused by beak trimming all birds more or less important than those of cannibalism in a few?

A second ethical approach (called deontology) is that there are things we should do, and things we should not do, regardless of good or bad consequences. This approach may suggest that an experiment causing severe suffering to a turkey is wrong, even if it is likely to result in production of a new medicine. The approach leads to the dual concepts of duties and rights. Regan (1983) is the best known advocate of the position that these concepts should include human duties to animals, and animal rights. Many religious people feel that we have a duty to care for God's kingdom, including his animals. Others take a similar position in relation to nature or the environment: that we ought to care for the environment, including animals.

Theories of duties and rights sound convincing, but leave us with many difficult decisions. Does a goose eating my crops have the right not to be killed? If I have the right to kill that goose, why not another in a medical experiment?

There are also other approaches to ethics, such as an emphasis on the person carrying out the action rather than the action itself or its consequences.

Different approaches may produce similar conclusions, e.g. all of these approaches have been used to argue that we should do more for animal welfare. However, some conclusions differ. Thus utilitarians argue that using animals for human benefit is generally permissible if the benefit outweighs any harm caused. Proponents of animal rights say that it is not, because animals have the right not to be harmed.

Most people do not follow Singer or Regan in attempting complete consistency in their ethical approach. In this, they have the support of philosophers such as Mary Midgley. Midgley (1986) describes the argument between utilitarians and animal rightists as a 'football match' and says (p. 195): 'The idea that morality could be reduced to a single basic form is a foolish one'. Most people are concerned with both the consequences of their actions and the actions themselves (Reiss and Straughan, 1996). Sandøe *et al.* (1997, p. 15) recognize this tendency:

> [A] hybrid view which is attractive to many people combines elements from utilitarianism and the animal rights view. One version of this would say that there are certain things that one may not do to animals, no matter how beneficial the consequences, for example causing the animals to experience intense suffering. As long as we abstain from these things we can, on this view, reason as a utilitarian would do. For example, killing of animals or causing them mild distress or inconvenience may be allowed if sufficiently good consequences follow.

Others come to a different balance of views. For example, Rollin (1995) suggests that Western society is moving towards acceptance of ideas associated with animal rights. People do not necessarily believe in animal rights as such, but an increasing number dislike the idea of animals being killed.

The fact that most people do not have rigid or even well-defined approaches to ethics allows change. People's attitudes to the welfare of poultry are therefore likely to change in response to changing influences such as their economic circumstances, the practices of the poultry industry and the information they receive about those practices.

7.5 Concepts of Welfare

Another reason for variation in attitudes to animals is that people do not all have the same concept of welfare (Tannenbaum, 1995). Indeed, one common tendency is to avoid any explicit consideration of welfare and settle instead for a list of factors regarded as important. Needless to say, such lists are quite variable. A cause for confusion is that they often include not only factors that affect welfare, but also the effects of those factors on the animals concerned. The Welfare Codes issued by the UK's Ministry of Agriculture, Fisheries and Food (MAFF) (1987b), for example, provided such a list in the preface until recently (Table 7.1). Among the first few items on the list, we may note that shelter and fresh water are factors that should be provided by people caring for poultry, but comfort and freedom of movement are descriptions of the birds' responses to their environment. This confusion is unhelpful, although of course it is important to identify both how to safeguard welfare and what constitutes it (see Section 7.7). Revised Welfare Codes are removing this confusion (Department of Environment, Food and Rural Affairs, 2002a).

Three concepts are common among animal-centred approaches to the question of what welfare actually is, and people may believe one of these or a mixture of two or three (Duncan and Fraser, 1997; Fraser *et al.*, 1997). First, animal welfare may concern feelings such as pleasure and suffering (Chapter 8). Secondly, it may concern health and fitness, so that problems such as disease and injury are the most important challenges to welfare (Chapter 9). Thirdly, welfare may concern the ability of animals to express their 'nature', e.g. by living in natural conditions. This third concept has received less attention than the first two, and is therefore discussed in the next section rather than in a separate chapter. The concepts can be summarized as emphasizing animal minds, bodies and natures, respectively (Fig. 7.3).

Once recognized, these concepts can be identified in other treatments of animal welfare, including in those that present lists of important factors (Table 7.1). For example, the Brambell Report (Her Majesty's Stationery Office, 1965) said that welfare is 'A wide term that embraces both the physical and mental well-being of the animal'. The Committee also stated that 'In principle we disapprove of a degree of confinement which necessarily frustrates most of the major activities which make up the animal's natural behaviour'. In other words, it acknowledged the importance of the three elements of body, mind and nature. These three elements can also be identified in the Five Freedoms listed by the UK's Farm Animal Welfare Council (1997) (Table 7.2; Section 7.7), which include feelings such as hunger, physical aspects such as injury and aspects of naturalness such as expression of normal behaviour.

One more important issue is the differences that exist between different species. It was mentioned in the previous section that according to Singer's interpretation of utilitarianism, which many people follow at least partly, consequences of our actions for animals should generally be given equal weight to those for humans. A common

Table 7.1. Factors relevant to welfare, listed in the preface to the Welfare Codes of the UK's Ministry of Agriculture, Fisheries and Food (1987b).

Comfort and shelter
Readily accessible fresh water and a diet to maintain the birds in full health and vigour
Freedom of movement
The company of other birds, particularly of like kind
The opportunity to exercise most normal patterns of behaviour
Light during the hours of daylight, and lighting readily available to enable the birds to be inspected at any time
Floors/perches which neither harm the birds, nor cause undue strain
The prevention, or rapid diagnosis and treatment, of vice, injury, parasitic infection and disease
The avoidance of unnecessary mutilation
Emergency arrangements to cover outbreaks of fire, the breakdown of essential mechanical services and the disruption of supplies

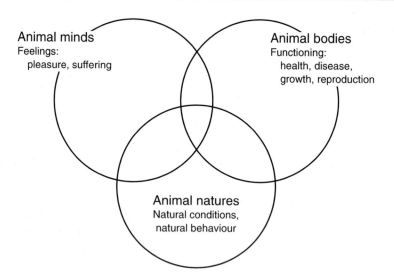

Fig. 7.3. Animal welfare: people's concepts of welfare emphasize animal minds, animal bodies, animal natures or a combination of these. The concepts overlap, but not completely (Appleby, 1999).

formulation of this is that there should be *equal consideration of similar interests* (Singer, 1975). This is sometimes misinterpreted to mean that, say, chickens should be treated like humans, with similar living conditions and other facilities, and the principle is therefore mocked and rejected. However, this is not what is meant. In fact, while there is considerable evidence that chickens have some interests in common with humans, such as not being in pain (Chapter 8), it is also obvious that humans are more complex than chickens, and have many interests that chickens do not. Much more complex conditions are therefore needed to safeguard human welfare than chicken welfare (Appleby, 1999). Nevertheless, it is now commonly argued that the conditions provided for chickens and other poultry in intensive production are inadequate for reasonable consideration of their interests, or for proper functioning of their bodies, or for expression of their natures.

The idea of expression of animal natures is explored in the next section.

Table 7.2. The Five Freedoms for animals recommended by the UK's Farm Animal Welfare Council (1997).

Freedom from hunger and thirst	By ready access to fresh water and a diet to maintain full health and vigour
Freedom from discomfort	By providing an appropriate environment, including shelter and a comfortable resting area
Freedom from pain, injury and disease	By prevention or rapid diagnosis and treatment
Freedom to express normal behaviour	By providing sufficient space, proper facilities and company of the animal's own kind
Freedom from fear and distress	By ensuring conditions and treatment which avoid mental suffering

7.6 Naturalness

One common complaint about battery cages, or broiler houses with no windows, or the fact that turkeys cannot mate on their own, is not completely covered by ideas of either physical or mental well-being: the complaint that 'it's not natural'. The philosopher Bernard Rollin (1993) suggests that this comes down to the idea of animal natures, a concept comparable with that of human nature:

> Animals have natures – the pigness of the pig, the cowness of the cow, 'fish gotta swim, birds gotta fly' – which are as essential to their well-being as speech and assembly are to us.

Concern for animal welfare then poses the question: what is necessary for animals to express their natures? There have been two main answers to this question. The first is that they must be kept in ways that allow them to perform natural behaviour. This is conveyed by Rollin's comment that 'birds gotta fly' and by Kiley-Worthington's (1989) claim that:

> If we believe in evolution, then in order to avoid suffering, it is necessary over a period of time for the animal to perform all the behaviours in its repertoire.

The second answer is that features of the natural environment such as sunshine and fresh air are important in themselves – a view held by the Swedish welfare campaigner Astrid Lindgren (Anonymous, 1989): 'Let the animals see the sun just once, get away from the murderous roar of the fans. Let them get to breathe fresh air for once, instead of manure gas.'

A problem with this approach to animal welfare is that it is difficult to translate into specific recommendations or requirements, and most legislation is specific. Is it really necessary for a hen to perform 'all the behaviours in its repertoire'? Take flying as an example. Domestic chickens rarely fly unless they have to, and it is easy to give them an environment in which they do not have to. Is it appropriate to enshrine in law a requirement for turkeys to 'see the sun'?

We cannot just say that animals should be kept in natural environments, for two reasons. First, natural environments are often impossible to characterize. What is the 'natural environment' of domestic chickens, after thousands of years of living with humans and genetic change from their wild ancestors? Secondly, life in the wild has many welfare problems, including bad weather, food shortage and predation. However, this may be missing the point. The point about respecting animal natures is not specific, and will not be addressed by drawing up another list of important factors. We can respect human nature without saying what sort of house people should live in or what food they should eat, and the same goes for poultry. It is not a question of whether turkeys specifically need to see the sun, but of treating them as turkeys, not as machines or humans or economic units. It is hard to translate this into practice, but progress is being made in some areas, including in the design of housing systems (Appleby and Waran, 1997). Some countries have also started to apply the concept of animal natures to legislation. Bernard Rollin (1995) points out that:

> Sweden has passed a law for agricultural animals that mandates that all systems of keeping farm animals must first and foremost accommodate the animals' natures. For example, the law grants cattle 'the right to graze' in perpetuity.

Putting this approach to welfare into practice would still need a lot of work. Allowing poultry access to the outdoors may solve some welfare problems but may be difficult in some environmental conditions, so it would be best to identify the features of the environment that are important for animals to express their natures. Excessive sunshine seems just as inappropriate for turkeys as constant lack of sunshine: perhaps it is variability that is important rather than sunshine as such.

As one of us has suggested elsewhere (Appleby, 1999, p. 36):

> Emphasis on animal natures will be particularly difficult to translate into detailed legislation. Yet it may still be useful as a corrective to the over-emphasis on specific details suggested by the other two approaches, a third leg to the animal welfare tripod of mind, body and nature.

So this volume will consider aspects of naturalness relevant to animal welfare, where appropriate.

7.7 Broad Approaches

A reasonable response to variation in attitudes to welfare, then, is to take a broad approach. Broad approaches may be divided into two categories: those that consider what welfare is and those that consider what to do about it. The first category includes the framework outlined above, that welfare concerns some or all of animal minds, bodies and natures (Fig. 7.3). It also includes one of the most widely quoted formulations relevant to animal welfare: the UK Farm Animal Welfare Council (FAWC)'s Five Freedoms (Table 7.2). These follow the common tendency, mentioned above, to address the question of what welfare is by listing factors regarded as important. The list is not definitive and there is no particular reason why it is grouped into five headings rather than four or six. It also suffers the limitations of a list approach, because the factors in the list may be contradictory. However, it is helpful, and will be used at intervals throughout this volume.

The current version of the Five Freedoms also includes a second column that illustrates the second type of broad approach: how to safeguard welfare. This has a lot in common with the factors relevant to welfare listed in the preface to the MAFF Welfare Codes (Table 7.1), which is not surprising as the Codes are mostly based on advice from FAWC to MAFF and to its successor the Department of Environment, Food and Rural Affairs (DEFRA). The MAFF/DEFRA approach to safeguarding welfare can be indicated further by the section headings in the Codes. Those in the Code for turkeys are given in Table 7.3 as an example.

The two approaches are, of course, complementary yet useful for different purposes. When the UK's Royal Society for the Prevention of Cruelty to Animals (RSPCA) launched its Freedom Food Scheme, the criteria it listed for farmers to sell their produce under the scheme were based on the Five Freedoms, because it wanted to make food production more animal centred (e.g. Royal Society for the Prevention of Cruelty to Animals, 1996). However, farmers found these criteria difficult to meet, because animals' responses to the environment are variable. They wanted to know what they were actually expected to do, e.g. what stocking density they should use. The RSPCA concluded that this was a reasonable expectation and

Table 7.3. Headings in the Welfare Code for turkeys of the UK's Ministry of Agriculture, Fisheries and Food (1987b).

Housing
  General
  Fire and other emergency precautions
  Ventilation and temperature
  Lighting
  Mechanical equipment and services
  Stocking rates
Feed and water
Management
  General
  Saddling of hens
  Toe cutting
  Beak trimming
  Desnooding
  Dewinging
Disposal of unwanted poults and hatchery waste
Handling and transport of stock on the premises

rapidly issued a completely new set of criteria on this basis (Royal Society for the Prevention of Cruelty to Animals, 1997).

These broad approaches express a considerable degree of consensus on the sort of factors that are important for welfare, even though there may be disagreement about specific details within those approaches. We are attempting a balanced approach to these issues in this volume that takes account of both the consensus that exists and such disagreements.

8   Sentience

8.1 Summary

- Sentience is the possession of senses, of feelings and of perceptions: all animals are sentient to a greater or lesser degree. There is broad agreement that welfare concerns sentience, and in particular the ability of animals to feel pain and suffer, or to feel pleasure.
- Studies of preferences ('asking the animal') and of needs can provide evidence of underlying motivation and thus of animals' feelings. Preference tests have shown that hens choose certain environmental features, such as cage size, floor type, nest site and food type, and select particular social factors.
- There is evidence that poultry possess cognitive representation; working to obtain a nest site even when it is not visible and even before they have experienced it.
- Situations where suffering may occur include deprivation of food and water when delivery systems break down or they are deliberately withheld, as in forced moulting or food and water restriction of broiler breeders.
- Pain and discomfort are caused by breast blisters, foot lesions, overheating, feather pecking, cannibalism, bone breakage, beak trimming and joint problems.
- Frustration and restriction of normal behaviour are more debatable, but may also cause suffering.
- Fear and distress can be associated with unfamiliarity, sudden or exaggerated environmental stimuli, handling or social factors, and are often worse in cages than in pens.
- Research in this area has identified many negative factors that can cause problems: it is less easy to determine if and when poultry are 'happy'.

8.2 Variation Between Species

The concept that welfare concerns the perceptions or feelings of animals is perhaps the most common of the three approaches to welfare outlined in Section 7.5 and Fig. 7.3. Dawkins, a foremost authority on welfare, has suggested (1988, p. 209) that 'To be concerned about animal welfare is to be concerned with the subjective feelings of animals, particularly the unpleasant subjective feelings of suffering and pain'. Duncan, who has written extensively on poultry welfare, goes further (1996):

> It is generally agreed that welfare is a term which cannot be applied sensibly to the lower animals or to plants but only to sentient animals. Since 'sentient' means capable of feelings, the argument is developed that welfare is solely dependent on what animals feel.

One reservation about this approach is that not everyone does agree: concepts of naturalness (Section 7.6) and of fitness and functioning (Chapter 9) may either be regarded as complementary or as alternatives. Verhoog and Visser (1997) suggest that it is being alive that is important: we should give moral consideration to all living things. However, most people would agree that feelings are relevant to welfare, even if they are not the only important factor.

The idea that welfare concerns feelings needs development, because 'feelings' may mean either just sensations such as touch and sight or more complex processes such as pain and emotions. This raises a problem, because whereas all animals have at least some of the five senses, it is less clear that all animals are capable of suffering or experiencing pleasure. The relevant question for this chapter is: can a bird suffer or feel pleasure?

Another word often used in this context is *sentience*, but this has the same spread of meanings as 'feelings'. Leahy (1991) says that 'To be sentient is to have the power of sense-perception; to see, hear, smell, taste or touch'. In this sense, all animals are sentient. However, the term may also be used to mean 'capacity for suffering or enjoyment' (Singer, 1975). It was in this sense that Duncan intended it, conveying the idea that some animals are capable of suffering while others are not, and he certainly includes poultry among sentient animals.

Resolution of the confusion comes from the fact that there is no rigid dividing line between animals with certain capabilities and those without. First, there is no sharp distinction between sensations such as touch and feelings such as pain. Secondly, all animals have mechanisms for responding to damage or avoiding potential damage. In vertebrates, these are similar to our own (Kestin, 1994; Gentle, 1997), but both the way in which the incoming messages are processed in the brain and the response to those messages differ between species. So chickens respond to stimuli that we call painful, but neither their mental nor their physical responses are necessarily the same as those of humans. It makes more sense to think of feelings such as pain being present to a greater or lesser extent in different species, than of them being simply present or absent. Thirdly, suffering is affected by thinking, and types of thinking vary between species. A particular animal species will therefore be able to suffer in certain ways but not others: chickens probably feel pain but not grief.

So we can say that all animals are sentient, but to varying degrees. The European Union (EU) has taken this approach recently, in declaring animals – without exclusions – to be 'sentient beings' (Ministry of Agriculture, Fisheries and Food, 1997). This suggests that welfare is a matter for concern in all animals, but that what this means in practice will depend on the degree of sentience of particular species, and indeed particular individuals. This supports the idea that poultry species can experience feelings such as pain, and emotions such as fear, but that the concept of welfare has little application to relatively simple animals such as the parasites of poultry that we wish to control.

We may also note briefly that on the basis of variety and complexity of behaviour shown, the mental processes of poultry are probably more complex than those of most fish, and comparable in at least some ways with both parrots (Pepperberg, 1987) and many mammals (Rogers, 1995).

There is one line of argument that produces the very different conclusion that welfare is only a matter for concern in humans, apes and possibly dolphins and elephants. This needs to be considered, but does not persuade many people. It is based on the idea that suffering requires a highly developed *consciousness*. Bermond (1997) argues that:

> Pain and suffering are conscious experiences. After all, it would be nonsensical to talk of experiences if those experiences failed to reach the domain of consciousness.

On the basis of experimental evidence and details of brain structure, he goes on to claim that only apes and possibly dolphins are conscious. As a result he says that:

> It is concluded that emotional experiences of animals, and therefore suffering, may only be expected in anthropoid apes and possibly dolphins.

However, Bermond is in a very small minority of opinion in two ways. First, he is using the term 'consciousness' in a specific way to mean 'the ability of an individual to perceive its own mental life' (Walker, 1988). This is more often called self-consciousness (or self-awareness), and most people would agree that self-consciousness occurs, if at all, only or mainly in apes. Dolphins and elephants have been less investigated but are other possible special cases. Most people use the term consciousness with a broader meaning: 'awareness of the environment, sentience' (Walker, 1988). So it makes more sense to consider consciousness, like sentience, not as all-or-nothing but as something that animals may possess to a greater or lesser extent. Secondly, few would agree that suffering requires self-consciousness. Most consider that many animals, including birds, can feel pain without the ability to think 'that hurt me'.

We may conclude again that suffering and pleasure occur to a greater or lesser extent in different species, and that different species may experience different kinds of feelings, but that it is not productive to attempt any categorization of animals into those that can suffer and those that cannot. Thus the feelings of poultry are not the same as those of humans, but are a valid matter of concern.

8.3 Preferences and Needs

One criticism of the concept that animal welfare concerns feelings is that we cannot ever know about those feelings for certain. However, we do have evidence, which can be used to draw conclusions just as a court uses evidence to produce a verdict even when there is no definite proof of guilt or innocence. This evidence is of two sorts: evidence about whether animals are happy (considered in the following sections), and evidence about what they want. These are not the same thing. If a person's or an animal's preferences are fulfilled, this may make them happier but it may not, so these approaches are partly distinct, but they are often not distinguished in discussions of animal feelings and welfare (Appleby and Sandøe, 2002). One reason why fulfillment of a preference may not make a bird happier, at least in the long term, is that there may be conflict between the short- and long-term consequences of its choices. For example, birds will generally choose a small food reward available immediately rather than a larger one that is delayed (Ainslie, 1975). This issue is exacerbated in artificial situations, where animals are presented with choices for which they have not been equipped by natural selection (Fraser and Matthews, 1997). A hen continues to enter a 'trap nest' to lay her egg even if on previous occasions she has had to remain in it for several hours until released by a handler (Duncan, 1978b). Hens attempting to choose a nest box from among many similar ones tend to investigate those already occupied, and sometimes become so gregarious in their nesting that they suffocate (Appleby *et al.*, 1984a). Nevertheless, it is clear that the preferences or choices of animals do provide evidence about animal feelings that can be taken alongside other forms of evidence.

Motivation was described in Chapter 3 as the stage in processing of information (e.g. in a stimulus–response sequence) that is perceived by the animal, and the same description can apply to other similar terms: preference, want, desire and choice. This helps to clarify three important issues about preferences.

First, a preference may be associated with a need that is critical for life – such as the need for food or the need to avoid a predator – or may have no such association. Yet this is irrelevant to the direct impact of that preference and its fulfillment or non-fulfillment on welfare. If a hen fails to find a suitable nest site, she will not die, but natural selection has produced strong motivation for conditions that would help the successful production of chicks, so when nesting motivation occurs it may be just as strong as that for food (Cooper and Appleby, 2003). Indeed, some needs do not have corresponding preferences. A high concentration of ammonia in the air is damaging to health, yet chickens show only a slight tendency to avoid it (Kristensen *et al.*, 2000), presumably because the need to do so did not occur in their evolutionary history. Life-sustaining needs will be considered in Chapter 9.

Secondly, the stimuli that provoke preferences may be very varied. And thirdly, what the appropriate response is – or indeed, whether a response is appropriate at all – is similarly varied. We shall consider different stimuli first, and then different responses.

The earliest approaches to considering birds' preferences for different stimuli involved offering them simultaneous choices between alternatives ('asking the animal'). It was shown that hens choose cage floors with a small mesh size, which give more effective support for their feet (Fig. 8.1; Hughes and Black, 1973), choose

Fig. 8.1. Hens prefer cage floors made of mesh with closely spaced wires rather than with larger spaces, probably because a small mesh gives better support for their feet (from Hughes and Black, 1973, with permission from Taylor & Francis Ltd).

a large rather than a small cage (Hughes, 1975a; Dawkins, 1981), choose enclosures that contain litter substrate rather than having a mesh floor (Hughes, 1976) and prefer to feed next to familiar cage mates rather than next to strange birds (Hughes, 1977). Preference tests also allow birds to be offered a choice between two or more desirable conditions. Hens show a strong preference for large cages as opposed to small cages when both have a mesh floor and for litter rather than mesh when the cages are the same size, but when obliged to choose between a small cage with litter and a large cage with a wire floor they prefer the former (Fig. 8.2). This suggests that they perceive litter as more important than space, at least in the context of this particular experiment (Dawkins, 1981).

Tests were then developed to assess the strength of preferences, concentrating on a single stimulus at a time. The most reliable assessment of motivation is obtained when several different variables all indicate its existence – in the case of hunger, for example, the amount eaten, the rate of eating, the willingness to tolerate an unpalatable substance such as quinine in the food and the amount of work that a bird will carry out to obtain food. This last method is used when birds are tested in Skinner boxes and have to peck at a disc for food, or when hens push through a door to reach food (Petherick and Rutter, 1990). In the same way, a number of separate tests have shown that access to a nest is important to hens: they are prepared to push through a weighted gate, pass through obstacles such as air blasts

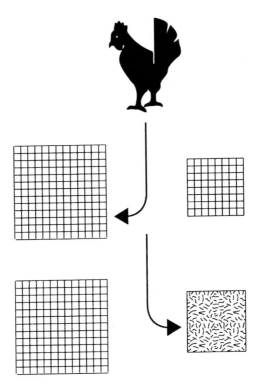

Fig. 8.2. Hens choose large rather than small cages, but prefer a small cage with litter to a large cage with a wire floor (from Dawkins, 1981, with permission from Taylor & Francis Ltd).

or pools of water, and walk long distances in order to reach a suitable nest site (Duncan and Hughes, 1988). A cockerel will work to reach a hen, but a hen will not work to reach a cockerel (Duncan and Kite, 1987). Tests have also shown that hens will strongly avoid certain stimuli, such as cage dusting – in which they respond to the approaching duster as if it were a predator (Rutter and Duncan, 1991). Another development has been what is called 'economic' analysis of motivation. If the work required for hens to reach food (the 'price' of food) is increased, hens work more to achieve the same intake: their demand for food is 'inelastic' (Petherick and Rutter, 1990) and food is therefore a 'necessity' (Dawkins, 1990). In contrast, one study of hens using litter for dust bathing found that they would not accept an increased 'price' for that opportunity and suggested that their demand for litter is 'elastic' (Faure and Lagadic, 1994), in which case economists would call it a 'luxury'. Evidence here is contradictory, however: another study suggested that demand for litter is also inelastic (Matthews *et al.*, 1993).

Consideration of preference for a single stimulus raises the question of whether birds would prefer such a stimulus to be present when it is absent, or if it is a case of 'out of sight, out of mind'. There is evidence that hens can have a cognitive representation of litter when it is out of sight (Petherick *et al.*, 1990) but controversy as to whether this means they are suffering in its absence: Widowski and Duncan (2000) conclude that they are probably not. However, Cooper and Appleby (1995) showed that hens would work very hard to reach a nest box on their first opportunity to do so. It has sometimes been suggested that for hens in cages that have never experienced such a nest site, 'they cannot miss what they do not know'. Cooper and Appleby suggested, on the contrary, that hens are strongly motivated to find an appropriate nest site – one that is enclosed, with a soft, level base – for their very first egg and every egg thereafter. Here there is overlap between the two approaches to animal feelings and welfare, the consideration of preferences and of suffering. We shall return to the latter below.

When we turn to the responses that birds make to stimuli, demonstrating a preference, the main question raised in relation to welfare is whether it is the outcome of the behaviour that is important to the bird or the behaviour itself. The UK's Brambell Committee (Her Majesty's Stationery Office, 1965) suggested that 'a very large part of animal behaviour is basically determined by innate abilities, proclivities and dispositions', and went on to argue that 'we must draw the line at conditions which completely suppress all or nearly all the natural, instinctive urges and behaviour patterns'. Animals are thus said to have 'behavioural needs'. Similarly, the Council of Europe's Convention on the Protection of Animals kept for Farming Purposes states that 'Animals shall be housed and provided with … care which … is appropriate to their physiological and ethological needs'.

The concept of behavioural or ethological needs was questioned by Baxter (1983) who supported a 'homeostatic' model of motivation in which the main function of behaviour was to return the animal to a desired state. He suggested that if all of an animal's basic needs are met, it should be unnecessary for any behaviour to occur. Subsequent reviews have emphasized, however, that the control of behaviour is complex and that there may be feedback from the behaviour itself as well as from the outcome (Jensen and Toates, 1993). Jensen and Toates therefore suggest that while behaviours themselves cannot be categorized into those that constitute needs and those that do not, there are instead needs associated with the performance of particular behaviours. To assess needs in this way requires that the animal be provided with the goal and then observed to see whether it still carries out the behaviour. This was done by Hughes *et al.* (1989) who presented hens with nests that they had constructed previously and found that they carried out just as much pre-laying behaviour as when they were given plain litter. If the behaviour is still seen when the goal is present, it strongly suggests the existence of behavioural need. Another related issue is that birds will sometimes carry out what seems to be unnecessary work: a tendency that is called contrafreeloading. Hens provided with freely available food in a Skinner box nevertheless pecked a key to gain access to identical food, and over a series of trials worked for an increasing proportion of their diet (Duncan and Hughes, 1972). This may be associated with a need for control over the environment, or a need for information (Inglis *et al.*, 1997), but in either case does seem to be a need or preference associated with performance of behaviour as suggested by Jensen and Toates (1993).

So animals may have preferences for certain changes in the environment, or for carrying out certain behaviour patterns. Frustrating those preferences may constitute a reduction in welfare, while fulfilling them may increase welfare, even if only temporarily.

8.4 Positive Feelings

The next consideration is feelings of pleasure and suffering, whether short or long term. There has often been an emphasis on negative feelings, and their avoidance, as in the quotation from Dawkins (1988) at the start of this chapter. Yet just as 'health … is more than the absence of disease' (Hughes and Curtis, 1997, p. 110), welfare is not just avoidance of the negative, but includes promotion of the positive (Mench, 1998).

Evidence about positive feelings of poultry is scant. One possible source is play. Animals that are playing look happy, and it is reasonable that play should be pleasurable to animals, as it is to humans. This is because play is functional, allowing animals to learn or practise actions that will be useful later (Bekoff, 1998), and natural selection may be expected to have produced positive reinforcement for such behaviour (Chapter 3). Young poultry of various species play occasionally, by jumping on and off objects, and chicks show 'popcorn behaviour', when running, jumping and flapping spreads rapidly through a flock. However, play is not as common or as varied as in many mammals, and is rare in older birds, as it is in older mammals.

Similarly, it is possible to interpret much normal behaviour as evidence of positive feelings. The idea that birds enjoy eating and that cockerels enjoy crowing was discussed in Chapter 3. Furthermore, the UK's Welfare Codes suggest that birds should be provided with comfort, and with 'the company of other birds, particularly of like kind' (Table 7.1; Ministry of Agriculture, Fisheries and Food, 1987b). Behaviour such as resting and affiliative behaviour (including flocking; Section 5.6) is therefore relevant here. Poultry generally rest in body contact with each other if possible, both on perches (Lill, 1968) and in 'rafts' on the ground. The functions of this behaviour must include thermoregulation and defence against predators and, as with play, natural selection may be expected to have made it positively reinforcing. It has also been suggested that hens dust bathe because the activity is pleasurable to them rather than being a behavioural need (Widowski and Duncan, 2000). However, this is obviously an area where there is a risk of uncritical anthropomorphism. It is easy to describe hens as showing 'contented clucking', more difficult to obtain good evidence of such contentment.

8.5 Hunger and Thirst

More evidence is available on negative feelings. Probably the most useful general word for such feelings is suffering, as already used extensively in this chapter. Some people would not describe mildly negative feelings as suffering, so use of the word may concentrate attention on more important, stronger feelings. Phrases such as 'pain and suffering' are sometimes used, as in the quotations from Dawkins (1988) and Bermond (1997) given above, but this is probably to give particular emphasis to pain rather than to distinguish it from other forms of suffering. There is no definitive list of forms of suffering; we shall use here those included in the Five

Freedoms (Table 7.2). This section and the next will consider those with direct physical causes: hunger, thirst, discomfort and pain.

It has been pointed out by Kyriazakis and Savory (1997, p. 50) that '"Freedom from hunger and thirst" is a *non sequitur*. Animals need to be hungry or thirsty in order to eat or drink.' So the emphasis here must be on severe hunger and thirst. These are not common in most poultry production systems, with certain important exceptions, because birds kept for eggs or meat are usually given food and water *ad libitum*. Furthermore, knowledge of nutritional requirements of poultry is more advanced than for any other category of livestock, so appetites for specific nutrients as well as general appetite will usually be met (Chapter 4). Poultry kept on smallholdings and left largely to find their own food may often be hungry. However, while hunger can result from either malnutrition or undernutrition, it is also affected by other circumstances: food-restricted animals that can 'do something about it' by foraging react much less than those in confined conditions (Duncan and Wood-Gush, 1972; Appleby and Lawrence, 1987). As such, even though poultry on smallholdings are often undernourished, hunger is probably not as critical an issue there as in the situations discussed next.

There are three main exceptions, in which severe hunger and thirst do occur in commercial poultry.

1. Automatic food and water supply systems sometimes break down. This is a particular problem in closed houses where the birds' daylight period is not aligned with real daytime and breakdowns may not be discovered for many hours. This applies to houses for laying hens, in which the light period often starts soon after natural midnight so that most eggs are laid before the operatives' working day. The effects of breakdowns are worst in laying cages and other simple housing systems where they cause more frustration than in more complex environments. Failsafe systems to cope with breakdowns are a priority, and indeed such systems are increasingly used.

2. Induced or forced moulting of laying hens is partly achieved by reducing day length but is considerably accelerated by methods that restrict eating, for example by providing hens with a novel food, which can cause them almost to stop eating for several days. The reason for this practice is that hens also stop laying, then after a break resume at a greater rate than before and lay eggs with stronger shells. However, these methods can cause hunger, and forced moulting is now rare in Europe. In the USA, moulting is usually induced by withholding food altogether for several days (Ruszler, 1998). This is now illegal in Europe and is increasingly controversial in the USA: in 2000, the McDonalds Corporation announced that it would no longer buy eggs from suppliers who induce moulting by withholding feed.

3. Broiler breeders are heavily food restricted because selection for the fast growth that is desired in their offspring means that if they are fed *ad libitum* they become obese and fertility drops. They are also water restricted because otherwise they drink so much that their droppings make their litter unmanageably wet. They frequently develop abnormal behaviour such as repeated pecking of spots on the pen walls, and experimental tests confirm that they are chronically, severely hungry (Savory *et al.*, 1993). Attempts to solve the problem, for example by offering bulky food or nutritionally modified diets, have not been very successful (Mench, 2002). It seems likely that this problem can only be significantly reduced by a radical change in the industry such as reversing the selection for growth rate in broilers (Appleby *et al.*, 1994) or use of dwarf parent strains.

8.6 Discomfort and Pain

Discomfort frequently occurs in association with other problems or as a lesser form of such problems. For example, meat turkeys often get breast blisters, partly because changed conformation results in the breast pressing on the feet when sitting (Wylie, 1999). Blisters are likely to be uncomfortable, particularly if they become infected. Similarly, discomfort is presumably a precursor to foot and claw damage of laying hens, which is common in cages. Thermal discomfort has been studied in broilers during transport. Mitchell and Kettlewell (1993) developed the index of Apparent Equivalent Temperature to express the interaction of temperature and relative humidity in causing heat stress, and this is now used in devices on transporters that indicate when minimal, moderate or severe stress will occur. A similar approach could be used in poultry housing in hot or cold climates, when overheating or chilling can be a major problem.

Pain in humans is associated with receptor cells called nociceptors, and birds also have these (Gentle, 1997). Poultry therefore suffer pain when they are subject to mutilations such as beak trimming, to feathers being pulled out by other birds, to long-term problems such as foot damage from wire flooring, to accidental injury such as bone breakage during handling or transport and to physical treatment such as hanging them on shackles for slaughter. Two situations where severe pain is likely have received particular attention. When beak trimming is carried out in mature birds, severed nerves attempting to regrow coil into masses called neuromas. These generate spontaneous nerve signals that in humans cause the stump pains after crude amputations. Trimming the beaks of adult birds therefore causes chronic pain in addition to the acute pain of the actual operation (Gentle, 1986b); this is uncommon in large-scale poultry production, but still practised in small-scale businesses. Trimming of chicks' and poults' beaks, while still controversial, does not cause neuromas (Hughes and Gentle, 1995).

Pain is also associated with joint problems and leg weakness in meat poultry. These problems have arisen largely as a consequence of selection for rapid growth, and affect a large proportion of broilers, as well as turkeys and broiler breeder males. Problems range from infectious disorders, such as necrosis of the femoral head, to developmental disorders such as bone deformity. The resulting gait impairment can make it difficult for the birds to walk to the feeders and waterers and is associated with pain (McGeowan *et al.*, 1999). Surveys of commercial broiler flocks indicate that approximately 30% of the birds have significant gait impairment (Kestin *et al.*, 1992; Sanotra, 1999). The industry has challenged this figure and has stated its intention of reducing leg problems by genetic selection. While some progress has been made, skeletal problems remain one of the most important and high-profile welfare problems of the poultry industry.

The occurrence of ascites in broilers is another example of a condition that causes both economic loss and suffering to the animal. First seen at high altitudes where atmospheric oxygen tension is lower, it occurs when growth is so rapid that tissue requirements for oxygen cannot be met by the cardiovascular system. Progressive right-sided heart failure leads to fluid extravasation into the pleural cavity, lung function fails and the bird essentially suffocates in its own plasma. In humans, it is a condition that causes terrible distress. The industry is tackling it by genetic selection for lines that have a better match between growth rate and cardiovascular development.

8.7 Frustration, Fear and Distress

There are other negative feelings that may not have direct physical causes, but are also aversive to birds and may thus constitute suffering, e.g. frustration, fear and distress.

The Five Freedoms (Table 7.2) include freedom to express normal behaviour, and poultry may be frustrated in this expression in various ways. Indeed, when hens are stocked at typical commercial densities in conventional laying cages, they are not afforded even an earlier, much more modest list of five freedoms. The Brambell Report (Her Majesty's Stationery Office, 1965) stated that 'An animal should at least have sufficient freedom of movement to be able without difficulty to turn round, groom itself, get up, lie down and stretch its limbs'. Dawkins and Hardie (1989) demonstrated that hens in laying cages do not have such freedom (Table 8.1). Furthermore, cages prevent or restrict pre-laying behaviour, comfort behaviour, feeding and foraging, and dust bathing. Inability to perform normal pre-laying behaviour (Chapter 6) is generally regarded as one of the most important problems for the welfare of hens in cages (Farm Animal Welfare Council, 1997). The reaction of at least some strains of hens indicates frustration (Wood-Gush, 1972). Comfort behaviours are also constrained in cages.

Restrictions on feeding and foraging in food-limited birds have been mentioned above. A restriction on caged hens, despite *ad libitum* feeding, is that although they generally strive to feed synchronously, they are prevented from doing so by typical cage widths of 10 cm or less per bird (Fig. 5.4). In the EU, there must be 10 cm of

Table 8.1. Area used by medium hybrid hens housed singly in small litter-floored pens (from Dawkins and Hardie, 1989).

Behaviour	Area (cm^2)	
	Mean	Range
Standing	475	428–592
Ground scratching	856	655–1217
Turning	1272	978–1626
Wing stretching	893	660–1476
Wing flapping	1876	1085–2606
Feather ruffling	873	609–1362
Preening	1151	800–1977

feeder per bird, but even less is common in the USA, and cages and feeders are usually the same width. Hens in cages are also prevented from showing full dust bathing, although they do carry out some of the movements on the wire. There is controversy about whether this causes frustration (Widowski and Duncan, 2000), but it does have physical effects, contributing to poor plumage condition.

All these behavioural restrictions are addressed by the European Directive on Laying Hens (Commission of the European Communities, 1999) which requires cages to be larger, higher and provided with perch, nest box and scratching area by 2012 (Chapter 13).

Fear and distress are probably associated with many of the other problems in this and the previous sections. They also occur independently. Hens in cages react adversely to approaching humans (Jones *et al.*, 1981) and to particular husbandry operations such as cage dusting, as mentioned above. In contrast, little or no avoidance of humans occurs in floor housing in at least some strains of hens: brown laying hens tend to cluster round people entering deep litter systems. There is other evidence that pen-housed birds are often less fearful than caged ones: birds in cages show marked fear responses to a novel stimulus, whereas similar groups in pens are indifferent (Hughes and Black, 1974). Another approach to assessing fearfulness is tonic immobility. It is well known that birds go into a trance-like state when turned upside down or treated in other, similar ways, and tests show that duration of this state correlates with other measures of fearfulness (Jones, 1986). Tonic immobility was found to be much longer for hens from cages than from pens (Jones and Faure, 1981).

In extreme cases, fearfulness results in hysteria: birds may injure themselves in cages (Rutter and Duncan, 1989) or pile up and suffocate in a pen. Hysteria is associated both with large group size, which allows positive feedback between birds, and with barren environments. In batteries, feedback occurs between cages, so group size in this respect is effectively large. In one series of experiments, varied attempts were made to reduce hysteria in colony cages. Some improvement was achieved by adding a tranquillizer to the diet, by claw trimming and by reduction of group size or stocking density. Complete prevention of hysteria, though, was only achieved by enrichment of the cage environment, by addition of nests or perches (Hansen, 1976).

Fear and distress are also caused by aggression. This occurs less often in cages than in other systems, probably because movement is restricted and subordinate birds are close to a dominant, which inhibits their aggression towards each other (Hughes and Wood-Gush, 1977). The 'peck order effect' (Duncan, 1978a) is also less likely in cages than in large groups, where it is common for a small number of low-ranking birds to be pecked continually by others.

We can also draw conclusions about fear and distress from physiological measurements. Duncan and colleagues (1986) measured the heart rate of broiler chickens that were being caught by hand or by machine. Heart rate went up in both groups, which suggested that the birds were frightened. It returned to normal more quickly in the latter group, so they concluded that it was less frightening for a chicken to be picked up by a machine than by a person.

Finally, Duncan and Mench (1993, p. 72) have raised the issue of other types of negative feelings:

Is it possible that poultry could experience other states of suffering such as loneliness, boredom or grief, or perhaps even states not experienced by human beings? There has been little investigation into these areas and so no hard conclusions can be drawn.

While that is true, we noted above that poultry do not show such complex cognition as higher mammals, so it is perhaps unlikely that they experience the more complex emotions that we do. Duncan and Mench (1993, p. 73) suggest that 'Considering the barrenness of many husbandry systems, boredom would seem to be a good candidate for further studies'. Can poultry be bored? By analogy with humans, boredom can be described as an unpleasant emotional state resulting from lack of general stimulation, once specific motivations are satisfied. In other words, boredom could only be confirmed in animals that were not motivated to feed, nest or carry out any other specific behaviour. If such animals had a residual need for stimulation, for example if they were motivated to explore, they could be said to be bored. There is little evidence on this question for poultry or any other animals. It is an important issue, though, because it affects whether husbandry systems which allow most or all specific motivations to be satisfied are appropriate, even if they involve confinement or other aspects which may appear unattractive to some observers. For example, the enriched cages allowed for in the European Directive (Commission of the European Communities, 1999) do not meet the requirement of some welfare groups that cages should be banned altogether; whether they are nevertheless satisfactory from the point of view of the birds' behaviour may depend on hens' capacity for boredom.

Even if boredom is not a major problem in poultry, behavioural restriction is still an important issue. We have seen that many intensive poultry systems do restrict behaviour. At the time when these systems were being developed, the effects of such restrictions were unknown. However, it is now clear that restriction of many behaviour patterns can cause frustration and that these specific frustrations amount to suffering. Other effects of the environment on behaviour can clearly also result in suffering, such as the influence of barren environments in producing hysteria.

To some extent, we can understand the feelings of poultry. We have made some progress in identifying problems, although less in being able to conclude that problems are absent and that the birds are happy. In the next chapter, we go on to consider the more measurable, physical aspects of welfare.

9 Physical Aspects

9.1 Summary

- Indicators of sound physical condition are necessary but not sufficient to show good welfare. Satisfactory production, a positive indicator, is not in itself proof of good welfare, whereas low production, a negative indicator, generally does seem to correlate with poor welfare.
- Though flocks are much larger, the incidence of infectious disease has declined greatly since the 1930s. This is because of changes in husbandry such as 'all in, all out' management, poultry health schemes, separating droppings from birds, cage housing, selection for disease resistance, antibiotics and vaccination.
- Some disease conditions, such as egg drop syndrome, have few adverse effects on welfare, whereas others with clear-cut and severe symptoms, such as *Escherichia coli* infection, probably cause considerable suffering.
- Injuries are also a problem: they include bone weakness and breakage, trapping, cuts and bruises, foot damage, hockburn and breast blisters. Causes include human handling and disturbance, cannibalism, accidents and poor design. Beak trimming, dubbing of combs and de-toeing also cause welfare problems.
- Healthy growth is both a commercial aim and, frequently, an indication of welfare. Problems arise when genetic selection has led to a growth rate so rapid that normal functioning is compromised, and also where food restriction is imposed to limit it.
- Farmers need to protect against foreseeable disasters, which can cause equipment breakdown, overheating, chilling, building failures and catastrophic mortality.
- Physiological stress is a useful concept best understood as the biological response to a stressor: if mild it is beneficial, if extreme it can lead to distress.
- Understanding the ways in which the integrity of the bodies of poultry is

threatened by disease, housing, husbandry, environment and management is an important element of welfare.

9.2 Positive and Negative Indicators

Most people agree that welfare concerns animal feelings. Many people, however, think that progress in understanding and judging feelings is not enough, as it will be a long time, if ever, before we know how an animal feels about being healthy or ill. Meanwhile, these people say, we must study health itself, and other aspects of how well the animal's body works or functions. Gonyou (1993, p. 43) puts it as follows:

> Although the animal's perception of its condition must serve as the basis for well-being, research in this area is only just beginning. At the present time much can be accomplished by using more traditional approaches involving behavioural, physiological and pathological studies.

Some go further. Hughes and Curtis (1997, p. 110) suggest that 'Many veterinarians [think] that taking care of an animal's physical health will automatically take care of its mental health'. Furthermore, an animal may have an illness of which it is unaware – which does not yet cause it to suffer – and some people think that this nevertheless reduces its welfare (Broom and Johnson, 1993).

Lastly, there are some people who do not apparently consider animal feelings at all. They often present their argument like this: if an animal is healthy, why should we worry about anything else? Those speaking for the agricultural industry have consistently argued that any system in which animals are physically healthy and in which production is good must, therefore, be a humane and satisfactory one. Such arguments, although not entirely without merit, are suggestive of special pleading, are based on narrow definitions of health and productivity, and tend to be framed from the viewpoint of the producer rather than from that of the animal. They have also proved unconvincing as far as public opinion is concerned.

One reason why assessment of welfare solely in physical terms is difficult is that some indicators, in some circumstances, are necessary but not sufficient to demonstrate that welfare is good. Indicators may also be contradictory. For example, it is sometimes argued that if hens are laying well their welfare must be acceptable. However, hens have been selected over many generations for the number of eggs they lay, and a hen has egg production as a priority that overrides many other aspects of her biology. She goes on producing eggs, which need a lot of calcium for the shells, even if her bones are weak from loss of calcium (Appleby *et al.*, 1992). So the fact that she is laying well may indicate that some factors relevant to welfare are satisfactory: she is probably well fed and healthy in most respects. However, she may nevertheless have weak bones and severe foot problems. Good egg production does not prove that her welfare is good overall: it is not a *positive* welfare indicator. On the other hand, egg production may be useful as *negative* evidence, evidence of poor welfare. If a hen from a strain selected for high egg production lays fewer eggs than usual or stops laying altogether, there is probably something wrong.

Using physical indicators also requires full knowledge of the circumstances. If the hen under consideration is not from a strain selected for egg production, she will lay fewer eggs even under ideal conditions.

Another important point is that, as pointed out in Chapter 7, farmers and other people involved with the economics of animal use may be expected to be concerned with the health and welfare of their animals, from enlightened self-interest. However, their primary concern is with group performance. From the animals' perspective, it is the individuals that matter.

9.3 Health and Disease

For either groups or individuals, there is no doubt that disease can cause major problems. As the scale of the poultry industry increased and flock sizes expanded during the 1920s and 1930s, progress was delayed by a sharp increase in incidence of conditions such as Marek's disease (Hewson, 1986), salmonellosis, avian tuberculosis and coccidiosis. In the mid 1930s, Marek's disease alone could result in an annual mortality of 20% or more, while pullorum disease and fowl typhoid, because of possible egg transmission, posed a threat to human health. Poultry health schemes with systematic testing and slaughter of infected birds, together with selection for disease-resistant lines, were partially successful in limiting the problem. Later, suitable antibiotics were discovered in the 1950s, and vaccines were developed on a commercial scale in the period from 1940 to 1970 against the remaining poultry diseases of importance such as Newcastle disease and infectious bronchitis (Biggs, 1990). Another major, relevant development for disease reduction in laying hens was the battery cage, which separated birds from their droppings. This made it possible to house birds in close proximity in very large flocks (see Section 11.5). There are several good texts available on poultry health and disease (Jordan, 1990; Pattison, 1993; Sainsbury, 2000). However, major disease problems still occur, such as the worldwide outbreaks of avian influenza in 2004.

Much can be inferred about the impact of a particular disease on welfare by examining its nature, development, course and effect on behaviour. Indeed, in addition to the other ways in which behaviour is important for welfare (see, for example, Chapter 8 and Section 9.5), behaviour is vital as an indicator of disease and health. Inactivity, drooping posture and failure to eat may all suggest that something is wrong and, conversely, normal behaviour is likely to indicate health, or at least that any disease is subclinical (Hart, 1988). Conclusions for welfare are partly based on associations between illness and suffering. However, recognition of suffering is difficult even in the case of pain, as Loeffler (1986, p. 49) points out:

> Clinical assessment of pain depends mainly on visual inspection; however, pain may be present which cannot be detected in this way. Acutely painful processes in the limbs or spine are relatively easy to diagnose because they disrupt locomotion. Severity of pain can be assessed by measuring the intensity and frequency of such signs as locomotor activity and posture ... Three categories of pain may be recognized but uncertainty may remain even after careful observation.

Such assessments inevitably involve presumptions that can never be fully tested, but it is important to note that conclusions about the impact of diseases on welfare are not solely based on their effects on feelings. As already pointed out, the fact that 'something is wrong' is sufficient reason for many people to try to put it right, irrespective of how the animal feels about it.

Curtis (1990) attempted to assess the welfare impact of specific disease conditions systematically, based on the duration of the condition, the nature of the lesions and the degree of apparent pain or distress evinced by the birds. Some were classified as having little adverse effect on welfare. Egg drop syndrome is a term used to describe a decrease in production by laying hens of up to 40%, caused by a virus. The hens are normally healthy in appearance or may appear just slightly depressed for about 48 h (McFerran and Stuart, 1990). However, egg production is unlikely to recover, so the flock is usually slaughtered. Sudden death syndrome in broilers and laying hens, and a similar condition in turkeys, is self-explanatory, with little or no preceding morbidity. Some conditions, such as fowl pox, mycoplasmosis and salmonellosis, are described by Curtis as having intermediate or variable effects on welfare. Lastly, some may cause considerable suffering, such as Gumboro disease, pasteurellosis and *Escherichia coli* infections. To the latter may be added the leg problems of broilers and turkeys mentioned in Section 8.6.

The emphasis of the poultry keeper on the group, rather than the individual, is demonstrated by the fact that if a small number of individuals is ill, it is unlikely to be economically worthwhile to treat them. It is usual just to cull them.

Having said that, the fact that disease is such a major potential problem has led to the industry turning from a defensive to a positive approach, building on the historical development of vaccines to impose complex and active programmes of health maintenance and monitoring. Another reason for this is the intense competition between rival commercial concerns within the industry, for example between breeding companies, in developing their own strains of bird. Not all the effects of this are genuine: competition may lead to under-reporting of problems and aspects of promotion that are essentially cosmetic, such as invention of the term livability to avoid mentioning mortality. Indeed, figures quoted for livability are massaged, because, although purporting to be percentages, they actually represent birds surviving out of every 102 – arising from the practice of delivering 2% more birds than are ordered, in case some die. Nevertheless, health programmes do have major positive effects.

Such programmes include an extremely systematic approach to vaccination and medication of birds, particularly preventative. There is also an economic reason for taking a preventative approach, which is that, if disease should occur, therapeutic medication is closely regulated to prevent contamination of human food by drug residues, so use of some drugs necessitates discarding eggs or carcasses. In addition, however, health maintenance involves many aspects of farm design (such as separation between buildings), biosecurity (for houses, workers, visitors and vehicles) and general management, as outlined in Table 9.1. Perhaps the most important element of general management is the 'all in, all out' approach developed by the poultry industry and only now being adopted, many years later, by other sectors of agriculture such as pig farming. The ability to clean facilities thoroughly between

Table 9.1. Health maintenance and monitoring: important areas to consider (adapted from Julian, 1995).

Housing and equipment
Environment and safety
Isolation and sanitation. Programme should include:
    One age group on farm or keep groups separate
    Obtain replacements from a single, disease-free source
    No neighbouring poultry within 300 m
    Clean and sanitize between crops
    Keep out wild birds, vermin, cats and dogs, and control insects
    Reduce movement between houses, change clothing and wash
    Use clean egg cases
    Employees should not keep or visit other poultry
    Restrict visitors and provide clothing
    Control vehicle movement
Nutrition and feeding
Immunization
Medication
Monitoring

flocks, and to reduce transmission of disease vectors from old to young animals, has made a paramount contribution to poultry health.

A final important procedure is monitoring (Julian, 1995). One aspect of this that is required by law in many countries is the daily inspection of all birds, and it is usual for workers to walk through all poultry houses at least once a day. However, in so far as the intention of the law is to enable identification of sick birds, it is, in effect, widely flouted. A worker walking between rows of battery cages six tiers high may be unable to do much more than spot dead birds, if that. It is slightly easier to see ill birds in floor-housed flocks, but even there subtle problems are unlikely to be noticed by workers who are primarily looking for dead birds, given the huge numbers of birds that are now managed by each worker.

Poultry industry literature and advertising promote images of healthy birds. In relation to success in overcoming disease, this is largely a fair impression. However, 'Health … is more than the absence of disease' (Hughes and Curtis, 1997, p. 110), and in relation to other physical aspects of welfare this positive approach has not been so thoroughly taken on board.

9.4 Injuries and Mutilations

Another important aspect of health is freedom from injury. Injuries are diverse and occur in diverse circumstances; we shall consider first injuries associated with the physical environment and handling, and then procedures carried out by humans, such as beak and toe trimming. Injuries are also caused by interaction between birds. Feather pecking, cannibalism and aggression were considered in Chapter 5, and there may sometimes be accidental injuries too. For example, lack of claw wear is a problem in broilers, caged hens and turkeys, and they may scratch each other with the resulting needle-sharp claws (Frankenhuis *et al.*, 1991).

It used to be common for birds in many husbandry systems to suffer injury or death because of design faults such as narrow gaps in which their feet, wings or head could be trapped. However, recognition of this problem and comparison between designs of laying cages by Tauson (1985) led to considerable reduction in such injuries in cages (Tauson, 1988). Indeed, a considerable amount is known about design and management to prevent other injuries (Curtis, 1983; Wathes and Charles, 1994), but this knowledge is not always put into practice because of economics. Hens kept on thin floor wire in laying cages develop foot damage (hyperkeratosis and fissuring). Damage is less on thick wire, or if cages are fitted with perches (Fig. 9.1), but these measures also cause more egg breakage (Carter, 1971; Duncan *et al.*, 1992). Overgrowth of claws is also a problem in cages; it can be prevented by a claw-shortening device such as that devised by Tauson (1986): abrasive tape on the surface below the feeder shortens the claws as hens scratch while feeding. The economic benefit of preventing claw overgrowth is insufficient for widespread adoption of such a device, but from 2003 it is mandatory in the EU. Foot and claw damage are severe when ducks are kept in cages, as in Taiwan.

Laying hens are prone to bone weakness because of the high metabolism of calcium for egg shell production. In non-cage systems, they sometimes break bones, e.g. by failing to jump gaps accurately (Broom, 1990; Gregory *et al.*, 1990). However,

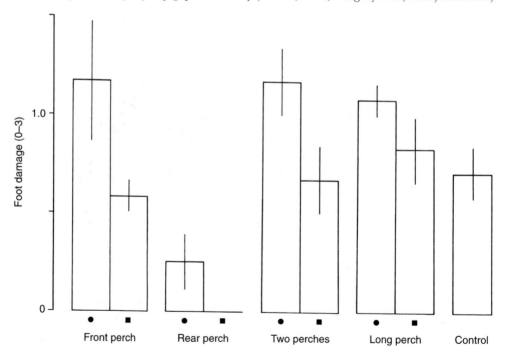

Fig. 9.1. Foot damage in laying hens at 66 weeks, in experimental cages with different arrangements of circular or rectangular perches, compared with controls. These cages had thick floor wire and damage was generally low, but a single, rear perch reduced damage further (from Duncan *et al.*, 1992, with permission from Taylor & Francis Ltd).

bone weakness is exacerbated by lack of exercise, and is therefore worse in cages by the end of lay. Up to 30% of caged birds suffer broken bones during catching and transportation, and more during processing (Gregory and Wilkins, 1989). There are around half as many such breakages in birds from free range or percheries (Gregory *et al.*, 1990). Bone strength can be increased by the addition of perches to cages (Fig. 9.2) and breakage reduced by improvement of cage fronts to make extraction of birds easier (Walker *et al.*, 1997), but perhaps the most important way to prevent breakage is gentle handling. Catching, transport and processing of broilers also frequently cause injury. One development that may reduce this is introduction of mechanized harvesting rather than catching by hand (Fig. 9.3).

Other improvements to laying cages that have reduced damage to hens include simplified cage fronts that reduce feather abrasion during feeding (Elson, 1988) and solid cage sides that also reduce feather damage (Tauson, 1989). Nevertheless, injuries still occur, for example because birds may react hysterically to the approach of workers, flapping frantically against the rear of the cage (Rutter and Duncan, 1989). In floor housing, hysteria can cause death if birds pile up. This can happen if workers only enter houses at long intervals, and birds subsequently panic. They can be accustomed to disturbance if workers enter houses and make a reasonable amount of noise on a systematic basis (see Section 10.9).

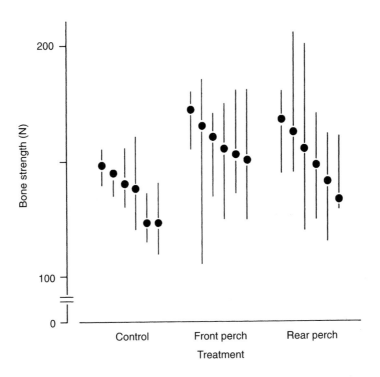

Fig. 9.2. Tibia strength in Newtons (N) of hens at end of lay, from cages with no perches, front perches or rear perches, measured by breaking the bones on a three-point rig. There were six cages in each treatment with four birds per cage; medians and ranges of values from each cage are shown (Hughes and Appleby, 1989).

Fig. 9.3. Mechanized broiler harvester. The broilers are drawn in by rotating rubber fingers and pushed on to a conveyor. (A) General view. (B) Pick-up head showing how rubber fingers interlock as they move inwards.

While accidental injuries to poultry of all species are common, it is striking and contentious that intentional, physical damage by humans is still more so, in the form of procedures such as beak trimming. Indeed, beak trimming of chickens (see Section 5.10) is almost ubiquitous except for broilers, and in Sweden and Switzerland where it is banned. Such operations are done with the intention of preventing more severe problems – beak trimming reduces feather pecking, cannibalism and damage from aggressive pecks – yet they are of increasing concern. Beak trimming (practised in most poultry species) causes pain and also removes the bird's second most important sense organ (after the eyes). The alternative approach of fitting 'spectacles' or 'bits' is common in pheasants. This may be less invasive – although some types penetrate the nasal septum – but also works by interfering with normal bodily functions: spectacles by blocking forward vision and bits by preventing full closure of the beak. Less common practices include dubbing (removal of the comb to prevent later damage from aggression or frostbite), desnooding of turkeys, and removal of breeding cockerels' inner toes to prevent damage to females' backs during mating. A list of operations sometimes carried out on ducks includes bill trimming, 'declawing, injecting, toe slitting or wing-tagging at day-old' (Gooderham,

1993, p. 262). All these must be painful at the time of the operation and can be criticized as unwarranted, invasive surgery, necessary only because of the circumstances in which we choose to keep these animals. Gooderham (1993) also mentions associated bacterial infections. The dilemma is real, but there is increasing pressure to use alternative solutions, either by changes in housing conditions or by genetics. Hope for a genetic solution to cannibalism is offered by the fact that some strains of chickens show less cannibalism than others and by experimental breeding programmes that have reduced this behaviour successfully (Craig and Muir, 1993; Muir, 1996). More attention should be given to this in future by breeding companies. Meanwhile, there are operations currently carried out that are excessive and, strictly speaking, unnecessary. Notably, laying hens are usually beak trimmed as chicks, removing about one-third of the beak. Yet 'beak tipping', removing only the very tip of the beak, is apparently just as effective in some strains and seems better for welfare; this is practised by some commercial companies and should replace more severe beak trimming wherever possible. Furthermore, many of these hens are destined for cages. As there are designs of cages available in which pecking problems are rare (see Section 12.6), and production is possible without this procedure in Sweden and Switzerland, beak trimming is increasingly unjustifiable.

9.5 Normal Behaviour

If animals are functioning properly, they are not just healthy. They are eating, drinking and excreting, moving, breathing and responding to stimuli (Chapter 4). Changes in any of these behaviours may indicate problems, and one advantage of automation is that some of these aspects of behaviour can be monitored indirectly. Thus food and water supply is often monitored house by house and, if there is a drop in consumption, i.e. decreased eating or drinking, the birds can be checked further. Workers may also notice other changes, such as reduced movement, panting or altered vocalizations, if they occur in a large number of birds.

Our understanding of normal behaviour comes from studying poultry carefully in a variety of environments and circumstances, as emphasized in Section 3.3 on the function of behaviour. In chickens, this has included study in feral conditions (Duncan *et al.*, 1978; Savory *et al.*, 1978). It was also pointed out in that section that differences in behaviour between environments are not in themselves evidence of malfunction, but that behaviour which causes problems for the individual concerned, and some other behaviour with no apparent function, such as stereotypic pacing, does indicate a failure to adapt to current conditions. Behaviour indicating problems for feelings, such as pain or frustration, was discussed in Chapter 8, and behaviour causing physical injury was discussed in Section 9.4. Stereotypic behaviour occurs in situations when a normal functional response of the animal is blocked, e.g. stereotypic pacing in caged hens unable to find or even seek properly for an appropriate nest site (Wood-Gush, 1972) and stereotypic spot pecking in broiler breeders on restricted food, unable to forage with any prospect of finding food (Savory *et al.*, 1992). In some circumstances, it has been suggested that stereotypies may actually be functional, perhaps reducing stress by causing release of morphine-like endorphins; individual broiler breeders that spot peck have lower corticosteroid levels than those that do not (Koštál *et al.*, 1992). However, it is better to avoid the stress than to reduce it later, and stereotypies have also been described as demonstrating a functional brain pathology (Dantzer, 1986).

Some aspects of behavioural expression have been affected by genetic selection, such as broodiness in chickens (see Section 3.4). Attempts are being made to reduce broodiness in turkeys using genetic engineering (Cochlan, 1993). Some ethicists consider this 'instrumental' alteration of animals inappropriate (Reiss and Straughan, 1996), but others disagree (Rollin, 1995).

9.6 Growth

In a biological context, physical functioning is associated with survival and repro-duction, as outlined in Section 3.3, so these aspects of life ought to be central to considerations of health and fitness. However, the picture is complicated by the ways in which poultry have been altered genetically, and by the ways in which they are treated for purposes of production. This also applies to a major contributor to both survival and reproduction, healthy growth.

Achieving healthy growth of stock is an important commercial aim. Formula-tion of diets to provide poultry with appropriate nutrition, to this end, is more advanced than for any other species, including humans. Yet this is not successful for all individuals. In particular, in most groups of poultry, there are a few birds that fail to thrive and remain considerably below average weight. Little attention has been given to this problem, other than to cull such runts as unlikely ever to do very well. Therefore, as with disease status, most consideration tends to be of group rather than individual performance. A widely used method for monitoring the growth of broilers and broiler breeders is an automatic, electronic weighing platform (Fig. 9.4). This has to have a lower limit for weights that it will accept – to prevent records from, say, birds with only one foot on the platform – and therefore tends to exclude runts.

One method used for many years for promoting growth, particularly in broilers, has been inclusion of antibiotics in feed. It appears that this worked partly by reducing competitive energy use of gut bacteria and partly by preventing low-level infections that, while not leading to overt illness, might otherwise have reduced growth. However, one UK broiler company recently announced that it had stopped this practice as trials had found it no longer effective; presumably, it had been obviated by many years of selection for food intake and growth rate. Another factor in their decision must have been the fact that the practice is increasingly controver-sial, for two reasons. First, there is increasing evidence that large-scale use of antibiotics in animal production is leading to emergence of antibiotic-resistant strains of bacteria, dangerous to both livestock and human health. As such, the EU is moving towards banning this practice altogether and the US authorities are requesting companies to stop using the most important human antibiotics for this purpose. Livestock producers are generally resisting these changes; one response to the controversy is that they no longer refer to antibiotics used in this way as 'growth promoters', but stress their role in protecting animal health. Secondly, however, this long-term, 'subtherapeutic' use of antibiotics in healthy animals is criticized by

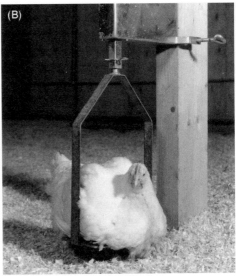

Fig. 9.4. Electronic perching platform for the automatic weighing of broiler chickens. Only one bird can stand on it at a time. (A) General view of young broilers. (B) Close up showing an older bird resting on the platform.

opponents of intensive farming as perpetuating unsatisfactory housing conditions: treating the symptoms rather than the causes of problems for animal health.

Growth is generally a negative indicator for welfare (see Section 9.2): below-average growth usually indicates a welfare problem, but there is no reason to suppose that above-average growth is associated with a welfare advantage. It is probably true that in birds not selected for growth, such as laying hens, growth on target shows that many aspects of welfare are satisfactory. The same may also be true for broilers and turkeys intensively selected for weight gain, but in these cases such positive aspects are offset by the problems associated with rapid growth (see Section 8.5). In terms of functioning, those problems are made even more obvious by recent findings that excessive growth of broiler muscle is associated with myopathy (Section 2.9; Sandercock *et al.*, 2001), as well as the well-known fact that growth of turkeys has been increased so much that they cannot effectively mate naturally.

The fact that growth less than an animal's potential indicates a welfare problem also applies – and very strongly so – to broiler breeders on restricted food, as emphasized in Section 8.5. The need for food restriction is an integral corollary of selection for growth in broilers – and realistic solutions are needed urgently. One obvious possibility for decreasing food restriction of broiler breeder females is increased use of dwarf strains (Appleby *et al.*, 1992).

There are, however, some situations where a degree of food restriction appears to be the best compromise currently available. This includes in young broilers, where a period of restriction, to an amount less than would meet the considerably increased appetite produced by artificial selection, reduces later leg problems.

9.7 Survival

As mentioned above, survival is an important aspect of physical welfare or functioning. However, it is important not to lose sight of the fact that no commercial poultry are allowed to live their natural life span of several years.

Of those birds that die before their allotted span, a minority nevertheless avoid significant welfare problems – those that die suddenly, for example from heart failure, or that are culled humanely for, say, being too small. In most cases, however, death is preceded by severe welfare problems. This includes deaths from all the diseases and injuries discussed above. The number of these is very variable, yet, as with failure to thrive, less emphasis generally is given to mortality in flocks than to other characteristics such as average productivity. This is presumably because flocks generally comprise very large numbers of birds, with each individual having low financial impact. However, a level of mortality that is regarded as satisfactory, such as 6% over a year in laying hens, actually involves the morbidity and death of a very large number of individuals and should not be regarded with complacency. Halving this level to 3% would represent a significant improvement in welfare, for the birds that do not die and probably also for others that suffer similar problems but to a less than fatal degree.

It is probably also reasonable that in considering the needs of birds in our care, we should place most emphasis on life-sustaining needs, as distinct from needs for health or comfort (Hurnik and Lehman, 1985, 1988), while noting that other needs may be perceived by a bird as just as important (see Section 8.3). Of course, poultry keepers generally do provide such life-sustaining needs, such as sufficient ventilation, food and water, appropriate temperature and protection from predators, because it is in their interest to do so. However, there are still instances where such provision breaks down and more could and should have been done to avoid this. Breakdown of food and water supply systems was mentioned in Section 8.3. Heat stress may occur even in temperate climates, yet cooling systems are not always adequate to prevent this. As one further example, in December 2000, broiler farms in Arkansas were damaged by ice storms, and as many as 500,000 birds were either killed by the houses collapsing or froze to death (Anonymous, 2001). The buildings were ill protected against such conditions, and no provision had been made for emergency slaughter of stock, despite similar damage in January 2000 and previous years (Martin, 2000). Disasters are unpredictable for individual farms, but happen predictably on a larger geographic and time scale, and should be planned for and guarded against.

9.8 Reproduction

Reproduction has been prioritized by both natural and artificial selection, so a reduction in reproduction tends to be another negative indicator of welfare, as already indicated. If a hen lays fewer eggs than expected, there must be significant problems. If she lays well, some aspects of her welfare must be satisfactory but others may be much less so. Indeed, some problems are caused directly by increased reproduction. As well as bone weakness, a large proportion of laying hens have tumours of the oviduct (Anjum *et al.*, 1989), caused by the high concentrations of sex hormones that now occur throughout the year instead of only in spring and summer.

Again there are complicated issues here. In broiler breeders, fertility is maintained by rigorous food restriction, as we have seen. Even more contentiously, commercial turkey reproduction is not by natural mating but by artificial insemination (Chapter 6). The ancestral lek system of wild turkeys did not lead to high fertility in captivity, and selection for body size reduced this fertility further.

The case of artificial insemination in turkeys illustrates a limitation to the usefulness of discussing biological functioning in relation to welfare. Within a rather limited conception of biological function, artificial insemination is advantageous to animals, in increasing their contributions to future generations above that which could be achieved by natural selection or even by artificial selection using natural mating. Further, there is a sense in which it is possible to think of the ancestors of domestic poultry as having made a successful 'choice'. This is because their descendants have been much more successful – at least in terms of numbers – than other individuals of the same species that remained wild. Chickens are by far the most numerous birds on earth. One controversial suggestion is that the concept of evolutionary fitness should be incorporated more fully into that of welfare, so that, for instance, a mother might improve her welfare by defending her young, even if she sustained injury in the process (Barnard and Hurst, 1996). Most people though, conceive welfare as relating more to individuals than to their genes. On this basis, it makes more sense to limit welfare considerations about functioning to physical rather than evolutionary aspects, as indicated in the title of this chapter. This is also more consistent with other approaches to welfare. In the case of the current example, artificial insemination obviously contravenes a conception of welfare as associated with naturalness (Section 7.6). Artificial insemination also involves restraining and manipulating the birds and, when repeated, may cause physical damage to the genital region; Druce (1993) has suggested that it causes negative feelings including distress, discomfort and probably pain.

9.9 Stress

Apart from tangible signs of things going wrong, one other major concept commonly used in relation to physical aspects of welfare or functioning is that of stress. Unfortunately the term has a confusing variety of uses. It sometimes refers to environmental factors that affect an animal, and sometimes to the effects on the animal, assumed to be harmful. In fact, a meaning somewhere in between those senses proves to be most useful. Moberg (2000, p. 1) suggests that stress is:

> The biological response elicited when an individual perceives a threat to its homeostasis. The threat is the 'stressor'. When the stress response truly threatens the animal's well-being, then the animal experiences 'distress'.

By this definition, not all stress is harmful. On the contrary, it is generally accepted that mild stress is beneficial, stimulating the animal's bodily systems and helping them to develop, and presumably also stimulating its mental processes.

Moberg (1999) presents a model of stress (Fig. 9.5) that makes clear both how this concept relates to biological functioning and the links between physical and mental aspects of welfare. Thus (p. 3):

> A stress response begins with the central nervous system perceiving a potential threat to homeostasis. Whether or not the stimulus is actually a threat is not important; it is only the perception of a threat that is critical.

As one example of this, exposure of free-range poultry to large areas with no cover leads to physical responses such as increased heart rate and behavioural responses such as avoidance, which can be interpreted as stressful in both functional and psychological terms. This is because natural selection has produced these measures as defence against aerial predators, and they still occur even if no such predators are actually encountered.

Excessive stress is sometimes deduced from behaviour or from physical effects such as failure to grow, but more often from physiological measurements, including

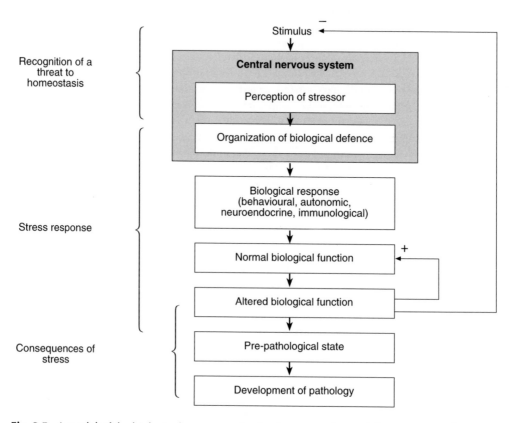

Fig. 9.5. A model of the biological response of animals to stress (from Moberg, 1999, with permission from Nature Publishing Group).

heart rate change, increase in corticoids and blood cell counts as classically described by Selye (1932). Physiological measurements of body function can therefore be made using heart rate monitors and blood sampling. However, as with physiological evidence of animal feelings, these are more often useful for comparisons between situations than for absolute judgements. Increased heart rate or corticoids do not themselves show that a situation is stressful: indeed, animals show the same symptoms during mating. However, there may be other reasons to assume that animals are stressed, including the 'threat to homeostasis' referred to by Moberg. Thus the heat stress studied by Mitchell and Kettlewell (1993) involves just such a threat. They used a number of physiological measurements to determine the effect of temperature and relative humidity on transported broilers (see also Section 8.5). Further, in this particular case, they were able to make absolute judgements of the severity of the stress, categorizing it as minimal, moderate or severe and helping to prevent the worst possible effect of severe stress, i.e. death.

In most cases, unfortunately, absolute judgements of the importance of physical measurements are impossible. We have also pointed out (Chapters 7 and 8) that many people find physical aspects to be inadequate for full consideration of welfare. Nevertheless, understanding the ways in which the integrity of the bodies of poultry is threatened by disease, housing, husbandry, environment and management is at least an important element of welfare.

10 Environmental Influences

10.1 Summary

- Poultry species are primarily ground-living. They are highly adaptable to a wide range of environments. Their tendency to seasonal breeding can be over-ridden by controlled lighting patterns.
- The main housing methods are caging and penning in large closed houses. Littered floors, which are widely used, and outdoor areas need careful management. The use of multiple tiers in alternative systems is increasing.
- Nutrition of poultry is arguably better understood than for any other species. They are normally fed a compounded diet that varies according to stage of development. Food restriction is used to keep broiler breeders healthy and in full production.
- Air quality is important, and airborne contaminants such as bacteria, dust and ammonia should be minimized. Relative humidity should be between 40 and 80%. Build-up of pathogens is reduced by 'all in, all out' management, by cleaning and disinfecting cages and pens, by restricting access of rodents and wild birds and, where used, by resting pasture.
- Birds can generally maintain body temperature (~42°C) over a wide ambient range. Egg production remains constant from about 15 to 27°C. In temperate countries, insulation and ventilation can usually maintain a suitable temperature without heating or cooling.
- Seasonal changes in physiology and behaviour are mainly governed by day length. Light control is used both to control maturity (short days) and to maintain egg production (long days). Intermittent light may be used to save electricity and to minimize food intake. Diurnal lighting changes are the main cue for egg laying.
- Welfare is affected by the interaction of social and physical factors: barriers

may curtail aggression, while multi-level houses can lead to fouling of birds but also aid roosting. Ideal conditions may be small groups at low density in complex environments.

- In most commercial systems, interaction with humans is negative: handling, catching and even inspecting can cause disturbance and fear. Good housing design is important in reducing this and also in making working conditions better for people.

10.2 Natural and Artificial Environments

Welfare is the outcome of the interaction between animals and their environment. In a commercial context, that environment has largely been chosen because of its effects on production, so this chapter will review environmental influences on behaviour and production as well as welfare.

The Order Galliformes, which includes pheasants, quail, turkeys and fowl, is predominantly a ground-living group. However, even within a restricted group such as jungle fowl, which includes the ancestors and relatives of domestic hens, there is considerable variation in both habitat and range. This reflects an adaptability that is an important pre-condition for domestication (Hale, 1962).

In dense cover such as scrub, forest or jungle, wild birds are difficult to study and there have been few detailed reports of jungle fowl or other galliforms in the wild. However, there have been two important studies of feral chickens, both of populations on islands, one in the Great Barrier Reef of Australia (McBride *et al.*, 1969) and the other off the west coast of Scotland (Duncan *et al.*, 1978; Wood-Gush *et al.*, 1978). These studies demonstrated how much the birds used cover, especially when roosting in trees or bushes at night. They often fed in open areas, and short vegetation was important for young chicks, allowing them freer movement. Their habitat was, therefore, complex, and their adaptability has been exploited during domestication.

Wild ancestors and relatives of domestic poultry include both temperate and tropical species. In common with other animals, the temperate species are more seasonal in their breeding, because chicks born in the autumn or winter would not survive. Seasonality is a disadvantage under domestic conditions, but the sensitivity of poultry to certain seasonal changes has been exploited to increase production. This has been achieved by controlling light conditions and will be considered further below. Controlled lighting is one aspect of the artificial environments that most poultry now experience, and the adaptability of these birds can be seen in their response to such environments. Birds can breed and lay successfully in a wide variety of artificial environments, from cages to pens containing three-dimensional arrangements of perches.

In artificial conditions, environmental influences can be divided into three categories that act separately and in combination: physical conditions, social conditions and effects of humans. As social conditions are considered in Chapters 5 and 6, we shall give most space here to physical conditions. In Section 10.8 we move on to the social environment and the ways in which this interacts with physical factors. We then consider contact with humans, an important part of the world of poultry, and, finally, the responses of birds to complex environments in which all these factors are simultaneously influential.

10.3 Structure

Small-scale, farmyard poultry systems are in some respects similar to the wild, with a small group of birds, often several females and one male. Laying hens are provided with a house and roosts (equivalent to bushes and trees), a reliable food source and nest boxes. Other poultry are given similar, simple facilities. There have been two main developments in larger systems: larger group size (in floor systems) or increased human control (in cages), or both. These developments have tended to involve housing, for protection and inspection of stock, for control of temperature and light and for reduction of labour. Decisions about housing depend on many factors, including climate. For example, closed houses enable the environment around the birds to be modified, which ordinarily results in increased food conversion efficiency among other effects. However, Curtis (1983), in the context of North American conditions, has suggested that in many cases totally closed houses are chosen primarily for labour-saving convenience and worker comfort, rather than for animal shelter.

The physical structure of floor or non-cage systems has many impacts on the behaviour, production and welfare of poultry. Relevant factors include whether birds have access to an outdoor area, and the nature of that area, flooring inside the house, use of multiple levels and facilities offered. The importance of different features of facilities is covered under the behaviour appropriate to them in Chapters 4–6, but it should be noted here that in commercial systems, features are often simplified or omitted, for both economy and control. For example, perches have been seen as unnecessary for hens, and water for swimming as unnecessary for ducks.

Outdoor areas generally offer varied stimuli, including supplementary sources of food or at least the potential of these, and hence increase the variety of behaviour shown. Use of such areas is uneven, affected by flock size and whether there is cover from aerial predators (Grigor *et al.*, 1995b). On balance, outdoor areas seem beneficial for welfare, but with free-range hens a regular topic of debate is the fact that not all individuals use them (Keeling *et al.*, 1988). Birds obviously need protection from predators and from infection by wild animals. If they are not shut outside, they usually avoid the most inclement conditions but can still have problems from mud. With large groups, birds tend to damage the ground in certain areas such as near the house entrances, so slats are often used. Inside houses, large groups also pose problems for hygiene and disposal of faeces. Since keeping the birds on wet, droppings-coated concrete would be harmful for their feet, slats, wire or litter are usual. Wholly slatted or wire-floored systems have been tried, but have severe problems: cannibalism is common unless birds are beak trimmed. Loose material allowing foraging is necessary to prevent this. The possibility of providing an area of sand or other loose material has received some investigation, but a simpler approach is to cover the floor with litter. The most usual litter is wood shavings, although straw is common for ducks, because it stays manageable while taking up their very liquid droppings. Hens use wood shavings heavily, for foraging and for other behaviour such as dust bathing, and also rest on them more readily than on slats or wire. However, maintenance of litter quality is important, ensuring that it remains friable rather than becoming wet or solidified. At its worst, wet litter containing a high proportion of faeces can become highly alkaline and cause hockburn and foot necrosis. This is a particular problem for densely stocked broiler chickens near the end of their housing period, as they produce large amounts of droppings and spend much time sitting or lying (Fig. 10.1). If litter is kept dry, though, by correct stocking and ventilation, bacterial breakdown of faeces occurs. One other problem for broilers and turkeys, if they are kept wholly on soft litter, is that their claws do not wear down and become needle sharp. The birds then scratch each other (Proudfoot and Hulan, 1985; Frankenhuis *et al.*, 1991) and the scratches often become infected.

Use of different levels for laying hens, in aviary or perchery systems, was first suggested by McBride (1970) as an evocation of natural habitats that include trees and bushes, although chickens in such habitats only use raised places for roosting. Once hens have learned to use different levels (Appleby *et al.*, 1988a), they will readily do so, and this produces at least some social advantages for welfare (see Section 10.8), e.g. allowing low-ranking birds to escape others. Physical effects, particularly on bone strength or breakage, are considered in Chapter 9.

Conventional cages for laying hens have pervasive problems for welfare, summarized in Table 10.1. There are also advantages: birds are in small groups and separated from their faeces, reducing disease and parasitic infections. However, in no other system are birds so intimately affected by every feature of the environment, which is completely artificial. Two points may be emphasized. First, hens spend a year or more, day and night, on a wire floor. The wire is thin, despite the fact that thicker wire would be less damaging to their feet, because thick wire has less 'give'

Fig. 10.1. Lesions such as hockburn and foot necrosis can be associated with the presence of wet litter (photograph courtesy of Lotta Berg).

Table 10.1. Some welfare problems caused by different characteristics of conventional laying cages (Appleby, 1993).

Characteristics	Welfare problems
Floor entirely of sloping wire	Foot and claw damage
Restricted area	Restriction of movement, causing bone weakness and breakage; restriction of specific behaviour patterns, some causing frustration
Undivided area	Prevention of escape from an aggressor or feather pecker
Restricted height	Frustration of comfort behaviour
Barren environment, no loose material	Frustration of dust bathing, foraging and pre-laying behaviour; claw damage; feather pecking

and eggs laid on it are more likely to crack (Carter, 1971). In Taiwan, ducks are also caged for egg production, which results in excessive growth of claws and ripping of webs, as well as very bedraggled feathers (Appleby, personal observation). Secondly, cage height is restricted. Nearly 25% of the head movements of an unrestricted hen occur above 40 cm (Fig. 10.2), but cages are usually 35–40 cm high. Quail are also often kept in cages that are too low; however, it is also deleterious to quail to keep them in cages that are too high, because they tend to jump as an escape response and severely damage their heads (Gerken and Mills, 1993). Other aspects of cages, and their modification to ameliorate welfare problems, are discussed in Chapter 12.

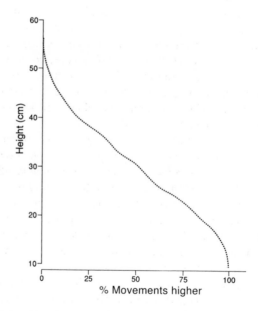

Fig. 10.2. The height of head movements by hens in cages of unrestricted height. Birds were filmed from the side, and each time a bird's head moved, the height it moved to was recorded (Dawkins, 1985).

One other aspect of the physical environment of poultry that has an impact on their behaviour and welfare, namely its complexity or, conversely, its simplicity or barrenness, is considered in Section 10.10.

10.4 Food and Water

The food and water available to poultry, including type, quantity and method of presentation, have major effects on their welfare. As we have emphasized (Chapter 4 and Section 8.5), more is known about nutrition of poultry than any other type of livestock and arguably than about any other species including humans (Classen and Stevens, 1995). This knowledge has been used in combination with genetic selection and other aspects of management to increase productivity, in ways that are ironically often detrimental to welfare. Thus, while most nutritional disorders in commercial poultry can now be avoided, maximization of production means that problems do arise. For example, maximization of egg production in laying hens and to a lesser extent in other species by breeding and feeding means that calcium metabolism is very finely balanced. Due to their high rate of egg laying, hens have osteoporosis which results in bone weakness. Other detrimental conditions may ensue if there are any problems with calcium supply, uptake or use.

Similarly, nutritional knowledge means that poultry are normally fed a single, compounded, nutritionally balanced food. Its composition varies only with the developmental stage that the bird has reached: 'starter' diets are high in protein and vitamins, 'grower' diets lower in both, and 'laying' diets high in calcium. Domestic poultry thus encounter food that is much less varied than that of wild birds and that takes much less time to eat. In addition, because it is usually provided *ad libitum* in the same location, it takes less time to find, and effects of reduced foraging and feeding time are discussed in Section 4.4. Again, knowledge of nutrition is part of the process that has led to rapidly growing broiler chickens and turkeys and hence to the need for restricting the food given to their parents. This restriction produces hunger and frustration (see Section 8.5). Various forms of restriction are used to limit growth and to improve food conversion efficiency, including skip-a-day feeding and limiting the quantity offered daily.

When food is restricted during the growth and production periods, the method by which it is supplied is especially important. Broiler breeders are generally fed by a feed line that runs several times per day. A certain length of feed line is provided per bird, but each time it begins to run, birds crowd round the point where food is appearing. More effort should be given to avoiding this, and indeed to addressing the whole issue of food restriction, as other problems arise from the need to restrict intake. Thus, one method of feeding females less than larger males is to put grids over the main feeders, too narrow for the males' heads, and to hang feeders for the males too high for the females to reach. However, the females are so eager to get food that they damage their eyes and heads on the grids (Duff *et al.*, 1989), and the males may also damage themselves trying to use the females' feeders.

Because a complete diet can be matched to only one particular level of production, it follows that birds producing at higher or lower levels will be underfed or overfed, respectively. One possibility is that this problem may be overcome by a self-selection dietary regime (Emmans, 1977). On small farms with deep litter or straw, this principle has sometimes been put into action by leaving part of the diet as whole grain, with the rest balanced as compounded mash. Scattering grain in litter also encourages foraging by the birds and improves the condition of the litter.

Wet mash is more palatable than dry, but has to be mixed fresh. This is demanding of labour and is now rare. When given only dry food, birds drink more than those in wild conditions, which eat a mixture of food items, many with a high water content. A reliable supply of fresh water will, of course, be particularly important in hot or dry conditions.

10.5 Air Quality and Hygiene

Air quality includes humidity, and hot air blowers may cause problems by drying the air. Relative humidity of up to about 60% is beneficial to growth in chicks (Sainsbury, 2000). Furthermore, respiratory infections are more likely in either dry air or very moist air, outside the range of about 40–80% humidity. High humidity can also cause birds to have difficulty keeping their body temperature down, because in hot conditions body heat is dispersed mainly by panting and evaporative cooling.

Respiratory infections are also increased by contaminants in the air, of which the most important are dust, bacteria and ammonia. Poultry breathing air with such contaminants develop lesions in the lungs (Maxwell *et al.*, 1989), which are associated with fluid accumulation and low blood oxygen (Odum *et al.*, 1987). These render birds more susceptible to infection (Anderson *et al.*, 1964; Oyetunde *et al.*, 1978). Problems with air quality are more common in floor systems than in cages, particularly where ventilation rates are low, although they may also be severe in cage houses with manure pits. In one study of a deep litter house stocked at low density, average airborne dust was 30 mg/m^3 and average ammonia was 23 ppm; the birds were exposed to these levels over long periods (Appleby *et al.*, 1988b). The recommended maxima in the UK for short-term exposure in humans are 10 mg/m^3 for dust and 35 ppm for ammonia (Health and Safety Executive, 1980). Ammonia levels can also become unacceptably high during the winter months in northern climates even in cage houses, because building heat is conserved by decreasing ventilation rates.

Air filtration is common in experimental houses but rare in commercial establishments. In an experimental study of its effects on broiler chickens, dust was halved and bacteria were almost eliminated (Carpenter *et al.*, 1986). Tunnel ventilation has some similar effects.

Although there have been good reviews of the disease problems which may be encountered with commercial poultry, such as that by Sainsbury (2000), there have been few systematic studies of the actual incidence of disease in different systems. However, the risk of diseases spread by contact between birds, or by contact between birds and faeces, is generally regarded as more severe in non-cage systems. Certainly, one of the main advantages of cages is that birds are separated from their faeces. This is important in the control of diseases such as coccidiosis. It is supported by 'all in, all out' management, with houses thoroughly cleaned between flocks (see also Section 9.3).

Similar management and cleaning are practised in other housing systems, but cleaning is, of course, not possible in systems incorporating pasture or dirt floors. In the former case, the usual practice is to alternate or regularly change the area of ground in use, to prevent build-up of disease. In most free-range systems, birds are also fed inside the house to reduce the risk of contamination by pathogens derived from wild animals. Some contamination may still occur on range or from wild animals entering the house. However, if the flock is initially disease free, husbandry is good, flock size is small and area of pasture per bird is large, then the danger of land becoming 'fowl sick' (infected) may be small (Hughes and Dun, 1986).

Metabolic diseases which are not spread by infection have been reported at a higher incidence in cages than in other systems (Duncan, 1978), while skeletal problems such as osteoporosis are most serious in the confined conditions of cages (Rowland *et al.*, 1972).

10.6 Temperature

As homeothermic animals, birds can maintain their body temperature over a wide range of ambient temperatures. In adult White Leghorn hens, this range is about -1 to $37°C$ (Esmay, 1978). Below this range, core temperature falls, while at higher temperatures it rises (Fig. 10.3); perhaps surprisingly, the upper end of this range is less than body temperature ($\sim42°C$ in adult hens), but $37°C$ is nevertheless higher than would normally be encountered for more than a few hours. Within this range is a narrower range in which metabolic heat production is at or close to a minimum: the thermoneutral zone or comfort zone. Thermoregulation in this range is physical, including behavioural. Its limits are usually called the lower and upper critical temperatures, but these are not distinctly defined and estimates of them vary. They are also affected by variation in humidity (van Kampen, 1981) and probably by acclimation to particular temperature conditions. The thermoneutral zone for hens is somewhere between 20 and $35°C$, probably slightly above that for humans, which is $22–30°C$ (Esmay, 1978). Yet even outside this range, there is no direct need for artificial temperature control, except in extreme conditions; with suitable diet, egg production is relatively constant across quite a wide range. This range has been estimated as from 15 to $27°C$ (Marsden *et al.*, 1987) or from 10 to $30°C$ (van Kampen, 1981). In temperate countries, the provision of relatively simple housing can prevent the birds from being exposed to extended periods of ambient temperatures outside at least the latter estimate of this range.

The main reason for close control of temperature is not productivity, but food consumption. In cold conditions, the need for increased heat production stimulates higher food (energy) intake, so it is economic to keep temperature in the thermoneutral range. For laying hens, $21–24°C$ is usually recommended. Above this, food conversion efficiency may be improved further, but egg weight may decline (Sainsbury, 2000) unless the concentration of nutrients in the diet is increased.

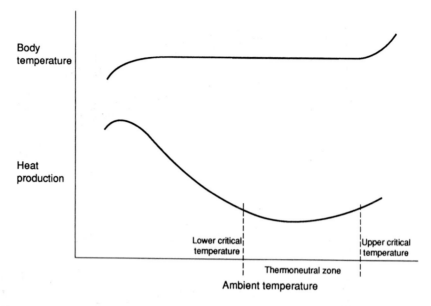

Fig. 10.3. Homeothermy. Below the thermoneutral zone (or comfort zone), the body burns energy to produce heat. Above this zone, it must also use energy to stay cool, and disperse the heat produced as well. At extreme ambient temperatures, the body's mechanisms for warming or cooling itself are insufficient and its temperature falls or rises.

There have been many experiments comparing performance of hens at different temperatures. One example is illustrated in Fig. 10.4.

For newly hatched chickens, the temperature below which they cannot main-tain homeothermy is about 26°C (Fig. 10.5), and is similar for poults and ducklings. Therefore, in normal, outdoor conditions, they need to be regularly brooded by the parent (Wood-Gush *et al.*, 1978). Artificial brooding therefore requires higher temperatures, of about 30–32°C, than are necessary for adults even though the body temperature of chicks at about 39°C is lower than that of adults. It is usual to decrease the brooding temperature gradually, rather than subjecting birds to a sudden change. For growing birds, either layers or broilers, a lower figure of around 21°C is the usual target. As with egg production, the reason is not that growth rates would decline below this but that food conversion efficiency would decline.

In temperate countries, house temperatures adequate for adult birds can usually be maintained by insulation and wind proofing, without the need for supplementary heating. This will depend to a large extent on stocking density. The number of birds in a house influences temperature directly, through the heat that they produce, and indirectly, through the amount of ventilation which is necessary. At low densities, heat production is low and in cold weather, in order to maintain an adequate ambient temperature, the ventilation rate has to be low. This may create problems with air quality and with damp litter. Higher stocking densities allow more ventilation, but very high densities may necessitate ventilation so great that tempera-ture control is difficult.

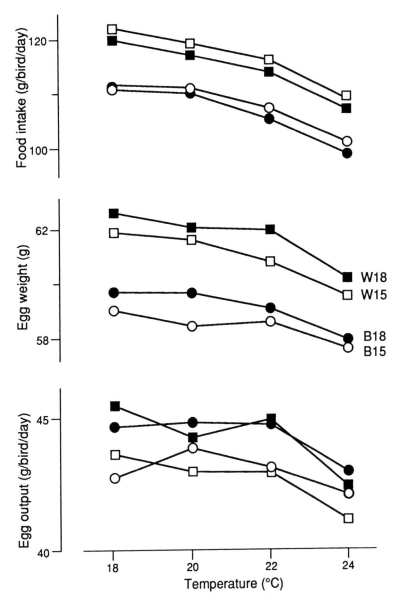

Fig. 10.4. Performance of hens at different temperatures. In this experiment, Warren (W) and Babcock (B) hens were fed on diets containing 18 or 15% protein (drawn from Table 3.5 of Emmans and Charles, 1977).

In countries with more extreme conditions, heating or cooling of houses, together with effective insulation, is frequently necessary. For best effect and most efficient use of energy, it is necessary to heat evenly the area occupied by the birds, which is most easily achieved by the use of hot air systems. Similarly, cooling systems that work on the air intake, such as foggers, misters and evaporative pads, are most effective at cooling the whole house. Other direct measures which are utilized to

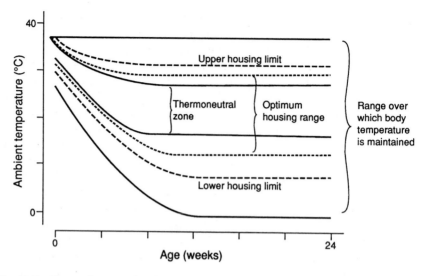

Fig. 10.5. Change in control of body temperature and in suitable housing temperatures with age, in White Leghorn chickens (after Esmay, 1978).

prevent overheating, including the use of open-sided houses, insulation of roofs to reduce downward radiation, the trickling of water over roofs exposed to the sun, and fans that increase air movement around the birds, are also very important. In fact, the practice of orienting buildings in relation to the sun and to other factors such as local winds, to prevent overheating, has probably been under-utilized in tropical conditions (Smith, 1981).

10.7 Light

The primary biological rhythms in poultry, as in other animals, are seasonal and diurnal, both mediated by light. The main factor controlling seasonal changes in physiology and behaviour is day length. This control is at least partly mediated by the hormone melatonin, which is produced by the pineal gland primarily during darkness: production is suppressed by neural signals from the retina resulting from incident light. The concentration of melatonin circulating therefore declines in spring and this has an effect on sexual development. For day length to have its controlling effect, it is necessary for the dark phase, or night, to be properly dark: light levels below 0.5 lux are recommended. To achieve this in housing when it is light outside, entry must be restricted to the light phase unless double doors are fitted; baffles are also needed around ventilators and other potential light sources. These measures, amounting almost to hermetic enclosure, are a major part of the artificiality or unnaturalness associated with commercial poultry housing.

Artificial control of day length for laying birds has two primary aims. First, it is an advantage for production to prevent birds maturing too early, at too low a body weight. Birds maturing early lay small eggs, not just initially but throughout their life. This is avoided by use of a constant but short day length during rearing, which has no obvious welfare implications. Secondly, light control is used to bring birds into breeding condition and to keep them in this state for an extended period. The welfare problems of this continuous reproduction are discussed in Section 9.8. It is achieved by a sharp increase in day length once the birds have reached their desired weight. In temperate countries, it is common to follow this by a slow increase (e.g. of 20 min per week) until a day length of 16 or 17 h is reached. Any advantage of this approach over maintaining a constant, fairly long day is, however, probably minor. In hot climates, the latter system is usually the only one possible, because in houses that are not fully enclosed the light period can be supplemented but not curtailed. Supplementing the natural daylight up to the longest day length of the year, or to a suitable longer period, avoids the problem of decreasing production as day length declines. A typical lighting programme for laying hens is shown in Fig. 10.6.

Despite the probable influence of melatonin in control of breeding condition, there is evidence for a direct effect of day length: of the separation of dawn and dusk as such. For example, 'dawn' can be indicated simply by a bright flash of light; if such a flash comes 1 h before the start of an 8 h light phase, this may have the same stimulating effect as 9 h continuous light. This discovery has been utilized in the development of intermittent lighting regimes (Fig. 10.7). In these, the light phase is interrupted by dark periods, which reduces electricity costs; and may increase food conversion efficiency without decreasing egg production (Rowland, 1985). Some of the regimes investigated have been complicated, but it seems likely that most of the benefit to be gained might be achieved on a simple programme, e.g. with one or two dark periods during the light phase. It has also been argued that an increased food intake during what would normally be a prolonged dark period improves egg shell quality by allowing more constant absorption of calcium. Intermittent lighting is also now used for meat birds. The main reasons are to reduce electricity costs, increase efficiency of food utilization and reduce skeletal problems. Thus, broiler chickens, which have usually been kept on a day length of 23 h or in continuous light to maximize food intake and growth rate, may grow as fast on an interrupted

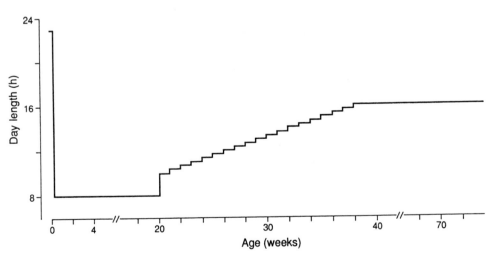

Fig. 10.6. A typical lighting programme for laying hens in an enclosed house.

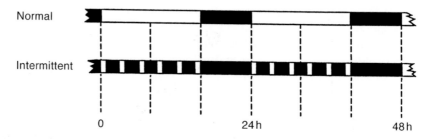

Fig. 10.7. Examples of different sorts of lighting regimes. In the intermittent regime, a light phase of normal length is interrupted by dark periods.

light schedule. Benefits in efficiency come from decreased activity during dark periods and perhaps from the reduction of 'boredom eating' in these birds, which have little else to do, resulting in improved digestion. It is not clear whether there any negative welfare implications of intermittent lighting. It is unnatural (as are 23 h day lengths or continuous light), but birds adapt to it readily, as they naturally alternate periods of activity and rest.

One other way in which day length is used to influence laying is in the induction of moulting (see Section 8.5).

In contrast to seasonal rhythms, diurnal rhythms do not necessarily involve day length. The main signal for controlling the pattern of behavioural and physiological variation over 24 h is usually dawn, although in continuous light or dark other factors can act as cues, such as feeding time, routine husbandry or temperature fluctuation. In fact, the main cue for ovulation and laying is dusk. In hens, most ovulation occurs during a period of about 5 h starting about 9 h after lights-out, i.e. around dawn on a typical light schedule. Laying follows about 24 h later (see Section 2.10) during the subsequent morning. The most common pattern is for an egg to be laid early in the day, then others at intervals of 24–28 h, with a fairly constant lag from day to day. A day is missed when early afternoon is reached, bringing the 'sequence' to an end; such a sequence is sometimes called a 'clutch', although there is little similarity between this and a natural clutch. After the missed day, laying begins early again. The length of sequences and the number of missed days determine the proportion of days on which birds lay. These relationships between the lighting pattern and laying are illustrated in Fig. 10.8.

Light intensity is also important for production. A certain minimum is necessary to stimulate ovarian function and so to maintain laying. This minimum is about 5 lux in hens (~1% of average sunshine), and intensities of 10 lux or more are usually recommended. Turkeys need brighter light, of 20 lux or more. However, bright light also has various effects on behaviour that are adverse for either the owner or the bird. It increases activity, and probably for this reason tends to decrease growth (Cherry and Barwick, 1962), because activity uses energy. It also increases aggression and feather pecking. For these reasons, the lowest practical intensities are used. Lighting for broilers may be as low as 0.2 lux, despite the facts that supervision of stock is difficult at less than about 1 lux and that such very dim light causes eye abnormalities (Manser, 1996). As with food restriction (Section 9.6), this is sometimes presented as a welfare dilemma, but it is a dilemma that the

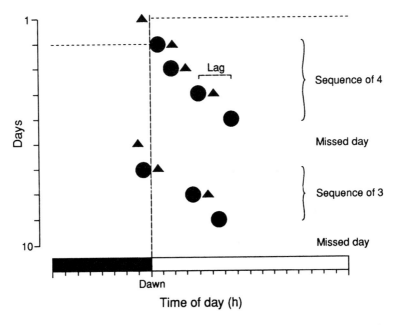

Fig. 10.8. Example of ovulation and laying times for one hen. Ovulation (triangles) is followed by laying (circles) after about 24 or 25 h, as indicated for the first egg. Laying is then followed shortly by ovulation of the next ovum, except for the last egg of a sequence: when laying occurs late in the day, it is not followed by ovulation so the next day is a missed one. In this example, seven eggs are laid in 10 days, so production is 70%.

industry has created by its emphasis on production, including its custom of keeping birds in large groups.

It is sometimes suggested that it is important for birds to have access to natural light. This may be reasonable as part of a general approach that natural conditions are more appropriate than unnatural. There is no evidence to support a benefit of natural light as such, but certainly some aspects of natural light variation help birds to adapt better to their environment. For example, gradual reduction of light at the end of the light period, rather than switching off lights abruptly, enables birds to feed in anticipation of the dark period, and to move to roost. In partly open houses, birds may react strongly to sunlight: if it shines only in restricted patches, they often crowd into these areas, even to the extent of piling on top of each other (Gibson *et al.*, 1985). Even in the absence of sunlight, it is generally important to avoid uneven light intensities in non-cage systems, as these result in uneven use of area.

10.8 Interaction of Social and Physical Factors

In considering environmental effects on the welfare of domestic birds, social factors have both direct and indirect impacts. First, other birds are themselves an important part of the surroundings. It has been said that 'a sheep on its own is not really a sheep' and poultry are similarly social. Individuals are rarely housed in complete isolation. Even when kept in single cages – as laying hens were when battery cages were first used and as all species of poultry are in some circumstances – they have social interactions with others through the wire, e.g. forming dominance relationships. However, it does seem inappropriate to keep social animals in social isolation even if keeping them in groups necessitates careful attention to avoid problems (Mendl and Newberry, 1997). When poultry are kept in groups, the influence of other birds may be negative, as in aggression, feather pecking and the fact that in multi-level systems birds may be fouled by others above them. However, social effects may also be positive, as in reproductive behaviour, mutual preening and the formation of groups that feed or dust bathe together. Social behaviour is considered in Chapters 5 and 6, together with the effects of group composition and group stability.

Direct impacts of group-mates on poultry welfare may also be physical, as when crowding leads to restriction of movement. Here the interaction between social and physical influences on welfare is most obvious, as crowding is caused by a combination of group size, total space allowance and stocking density (Appleby, 2004). The structure of the environment is also important, e.g. barriers, behind which birds can shelter, affect their use of the area (Newberry and Shackleton, 1997) and can curtail aggressive interactions.

Secondly, other birds have indirect impacts on welfare by modifying the physical environment, affecting such factors as temperature, air quality, litter quality and disease risk. There are also interactions between direct and indirect impacts, as for example when birds roost close together for warmth.

To emphasize the importance of both physical and social aspects of the environment, we have suggested that the best conditions for welfare are likely to be achieved in small groups of birds at low stocking density in complex environments (Fig. 10.9). Not surprisingly, that description fits natural conditions well.

10.9 Interaction with Humans

One aspect of the environment that was not part of the natural conditions in which poultry evolved, however, is humans. In addition to choosing the other aspects of the birds' environment, and carrying out all the husbandry practices (such as beak trimming, handling and artificial insemination) covered in other chapters, people have a considerable direct impact on welfare. In rare cases, this impact can be positive. If just a few birds are kept, they may come to regard their keeper as a social companion and react positively to the person's presence even without food or other incentives. However, probably more important for commercial poultry is the distinction between neutral and negative responses, in terms of whether the birds habituate to humans or react to them as predators. This depends both on the system in which birds are kept and the way in which the people behave. The extreme reaction of hysteria is more common in cages than in floor housing, with caged birds flapping violently and sometimes injuring themselves. This is probably because birds in cages see people more rarely than those in flocks, and the people they see are much closer. This interpretation is supported by the fact that birds in the top tier of cages, that have a poorer view of people, habituate less and are more fearful than

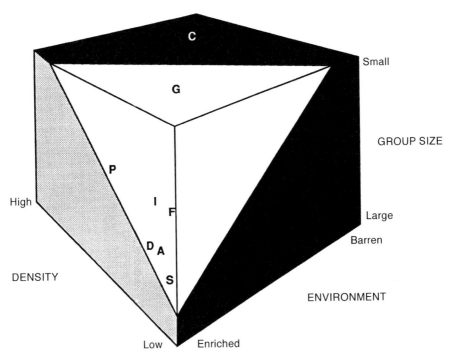

Fig. 10.9. Aspects of the environment interact in their effects on welfare. Here Appleby and Hughes (1991) suggested that welfare, particularly in relation to behaviour, might be regarded as satisfactory in the clear part of the volume, increasing further with increasing distance above the interface. Placement of systems on the diagram was tentative. A, aviary; C, cages; D, deep litter; F, free range; G, get-away cages; I, semi-intensive; P, perchery; S, strawyard.

those in other tiers (Jones, 1987a). When hysteria does occur in non-cage systems, however, the effects can be very severe, with birds piling up and suffocating each other. Hysteria can be prevented by moving among birds fairly frequently so that they become accustomed to people. This is – perhaps contrary to expectation – better than leaving the birds undisturbed, because they cannot be left undisturbed forever. It is also helpful to warn birds that someone is approaching, for example by knocking on the door before entering a house, rather than arriving suddenly.

Hughes *et al.* (1986) obtained evidence of the stressfulness of unaccustomed disturbance, even in the absence of hysteria. Handling laying hens that were housed on the floor, or even just passing through the pen, disrupted the process of egg formation and caused more abnormal eggs to be laid. Clearly it would have been better to accustom these birds to people before they started laying.

It is also important to note that several aspects of the environment discussed in this chapter are important for people working in poultry systems as well as for the poultry that they house. In particular, high concentrations of ammonia and dust in the air are unpleasant and potentially harmful, as indicated by the specification of maximum recommended exposure levels by the UK's Health and Safety Executive (1980). There are two other major aspects of system design with impact on workers.

First, ergonomics: the extent to which a system is comfortable and efficient to operate. Secondly, the general nature of the system: thus, many people work in the poultry industry because they like working with animals, and they may therefore dislike intensive methods such as the use of battery cages. Conversely, some more extensive systems are unpleasant to work in; for example, some aviaries for laying hens have platforms beside passages at head height, which allow birds to peck workers on the head. Failure to take these factors into account in designing systems has frequently led to frustration, low morale and high turnover in staff. This is a major matter for concern in itself. It also has considerable impact on poultry welfare. If it is physically difficult to inspect birds in cages, birds are unlikely to be inspected properly. If a deep litter house has a high ammonia concentration, workers do not spend any longer than they have to inside. If workers are impatient and frustrated, they are less likely to treat birds gently. Systems must be designed for humans as well as poultry.

10.10 Complex Environments

While it is useful to categorize aspects of the environment that affect welfare – as we have done here into physical, social and human influences – those aspects rarely act in isolation. Poultry react to the whole environment. For example, we have emphasized the importance of feeding for welfare, and feeding is affected by the physical nature of the food, the social interactions between birds and the way in which humans manage both food and birds. Two factors that are important in feeding are control and predictability. It is generally an advantage for birds to have control over their lives. A hen in a farmyard flock may get even less to eat, relative to her needs, than a commercial broiler breeder, but still does not show stereotypic behaviour. This is because she can do something constructive about her hunger: she can search for food with some prospect of finding it (see Section 8.5). Predictability of feeding is also particularly important when food is restricted. Broiler breeders fed on a regular schedule show considerable agitation if a meal does not arrive at its usual time. Considerations such as these remind us that a full assessment of welfare involves the environment as a whole, including system design, management and husbandry.

Indeed, in other circumstances, predictability is disadvantageous. In some aspects of the environment, variability is preferable, as when birds are being accustomed to disturbance. We may recall that variability is a major feature of natural environments, which many people consider to be essential to welfare (Section 7.6). Certainly there is evidence that environmental complexity is valuable. As mentioned in Section 3.5, chicks and quail reared with a range of novel objects and stimuli are less fearful and more able to resist stress later in life than those reared in barren conditions (Jones, 1982; Jones *et al.*, 1991).

A concept that overlaps with environmental complexity is that of environmental enrichment. The concept is less than wholly satisfactory, for two reasons. First, it often starts from an extremely impoverished basis. When birds are being kept in cages too small for them to stretch their wings, adding coloured plastic keys for them to peck (Gvaryahu *et al.*, 1994) may be beneficial but could hardly be described as giving them a rich environment. Secondly, the phrase more often refers to attempts to enrich the environment than to success. Thus, most benefits of plastic keys and other similar 'toys' are minor and temporary, as birds soon habituate to them. Part of the problem is that the changes made are often irrelevant to the animals (Newberry, 1995). However, if these strictures are borne in mind, attempts at environmental enrichment are more likely to be beneficial than otherwise.

Poultry are adaptable to different environments, but there are limits to that adaptability. Some have been considered above, e.g. susceptibility to extremes of heat and cold. Others concerned with particular aspects of behaviour are discussed elsewhere in this volume and many of these also have environmental components. An example is that sometimes hens start eating their eggs. The causes are unclear, but environmental factors such as high light intensity and crowding contribute to the likelihood of egg eating occurring. Another example is feather pecking, which tends to be worse in barren conditions at high stocking density. These have been cited as examples of birds failing to adapt to their environment, which raises the question of what is meant by adaptability. Birds in natural conditions do not eat their own eggs or peck other birds excessively. However, birds that show such behaviour in artificial conditions are adapting, in that they are behaving in a way advantageous to themselves. Their behaviour is not advantageous to producers, but birds do not behave altruistically for the benefit of producers. The challenge is to design production systems in which birds acting for their own advantage will also be acting for the advantage of the producers.

11   Industry

11.1 Summary

- Intensive husbandry depends on scientific and technological developments, including advances in incubation, nutrition, health, housing design, control systems and automation. Intensification was speeded by development of antibiotics, vaccination and 'all in, all out' regimes that improved poultry health.
- Enclosing houses reduces production costs by reducing food intake. Automation of food and water supply, manure and egg collection also reduces costs by reducing labour requirements.
- The largest poultry sectors are chicken meat and egg production, followed by ducks and then turkeys. Much of this output comes from large, integrated companies, comprising feed mills, breeding, hatchery, production, processing and marketing segments. Economies of scale reduce costs, but also foster a tendency to regard the birds as production units. Hatcheries are automated, but in the egg industry many surplus chicks are killed and not all countries use humane methods.
- Most laying hens are housed in cages. By about 72 weeks of age, shell quality and egg production have declined and hens are either moulted or culled. Moulting usually involves feed withdrawal and is rare in the EU but common in the USA.
- Broiler strains grow very rapidly and are normally housed in littered pens in flocks of up to 20,000. Welfare concerns include high stocking density, dim lighting, poor mobility, breast blisters, leg problems and ascites. Turkeys and ducks are often housed under similar conditions.
- The breeding sector is less automated, with birds kept as selected grandparent and parent stock in pens with natural mating (except for turkeys). Males and females receive different diets. Genetic selection, hitherto mainly for production traits, could make a contribution to reducing welfare problems.
- Processing is largely automated. Stress can occur during catching and loading,

transport and slaughter (due to failure to stun, causing birds to enter scalding tanks alive). Research is leading to better catching and handling techniques, improved transport vehicles and novel killing methods such as gas stunning.

- Welfare regulations and guidelines are increasingly prevalent in areas such as stocking density, beak and toe trimming, feed withdrawal, dust and ammonia control, monitoring automated processes and improving handling and slaughter. There is awareness of the need for first class management and husbandry.

11.2 Historical Development

The modern poultry industry originated in the 19th century, though its roots, as described in Chapter 1, go back to Roman times. It developed progressively from numerous small flocks of dual-purpose breeds that were fed on scraps and homegrown grains, and raised using natural lighting, incubation and brooding.

Along the road to large-scale intensive husbandry there have been a number of important scientific and technological developments, which have allowed almost all of the factors important for economic production to be brought under control. These included the artificial incubator and an effective brooding system, increased understanding of nutritional requirements and formulation of balanced diets, reduction and elimination of disease through improved hygiene, disinfection, vaccination and therapeutics, environmental and photoperiodic manipulation through housing design and the availability of electrical power and accurate control systems, and finally reduction of labour requirements by automation of husbandry procedures such as provision of feed and water, egg collection and manure removal. Of all livestock enterprises, poultry production is the most dependent on scientific knowledge and technical expertise. In the next six sections, we describe the most important scientific and technical advances.

11.3 Control of Incubation and Brooding

One of the earliest scientific developments, which enabled the industry to provide large numbers of day-old chicks whenever required, was the modern artificial incubator, which resulted from the invention of a simple and accurate thermostat in 1881 (Hewson, 1986), combined with gas-fired heating, effective control of humidity and a simple automated method for turning large numbers of eggs. Once the eggs had hatched, gas-heated brooders could be used to keep the chicks warm. It took many years of development before artificial incubators were as successful, in terms of hatchability, as broody hens. Indeed, bantam hens have been used to hatch small numbers of pheasant eggs almost up to the present day, and are still used to hatch the eggs of some endangered species of birds.

11.4 Control of Nutrition

Developments in the scientific understanding of nutrition started in the late 19th century. Initially all poultry rations were home-mixed, but as the potential egg output of laying hens increased and the essential role of proteins, vitamins and minerals became better understood, a complete, balanced diet could be formulated. This meant that birds no longer had to be given access to pasture and herbage in order to forage for themselves. The discovery in the 1920s that vitamin D could be given to chickens by adding cod liver oil to the feed paved the way for intensification of production, since it meant that birds no longer had to be exposed to natural light to synthesize vitamin D and could instead be housed completely indoors, allowing a much higher level of environmental control. By the 1930s, a number of commercial firms had begun to supply compounded feedstuffs in mash or pelleted form suitable for high-producing laying hens (Hewson, 1986). Diets were later tailored to meet particular needs: low-energy high-protein diets for growing birds, high-energy lower-protein diets for light hybrid layers, and very high-energy high-protein diets for broilers. One very important development was the discovery in the 1950s of methods for synthesizing essential amino acids, especially methionine, which meant that grain-based diets could be supplemented to make them nutritionally adequate.

11.5 Control of Disease

As the scale of the industry increased and flock sizes expanded during the 1920s and 1930s, intensification was delayed by a sharp increase in the incidence of conditions such as Marek's disease (Hewson, 1986), salmonellosis, avian tuberculosis and coccidiosis. Mortality rates in flocks rose from the 5–6% typical in earlier years to 20% or more (Smith and Daniel, 1982). Because of possible egg transmission, pullorum disease and fowl typhoid posed a threat to human health. Poultry health programmes involving systematic testing and slaughter of infected birds, together with selection for disease-resistant lines, were partially successful in limiting the problem. Suitable antibiotics were discovered in the 1950s, and vaccines against the remaining poultry diseases of economic importance, such as Newcastle disease and infectious bronchitis, were developed on a commercial scale in the period from 1940 to 1970 (Biggs, 1990).

   With this development, and with the separation of laying hens from their droppings by putting them in battery cages, it became possible to house birds in close proximity in very large flocks. As late as 1967, the average flock size in the UK of cage-housed laying hens was only 2200 (Sainsbury, 1971). By 2000, 97% of hens were kept in flocks of 20,000 or more, with 65% in flocks of 100,000 or more (Department of Environment, Food and Rural Affairs, 2000). The trend has been similar in the USA, and the majority of houses now have a capacity of from 30,000 to more than 200,000 hens (National Animal Health Monitoring System, 2000).

   Vaccinations used to be given by subcutaneous injection, but in large hatcheries vaccinations for many common poultry diseases are now given before the chicks are even hatched, using an automated process that injects the vaccine through the egg into the embryo. Other vaccines are sprayed on recently hatched chicks, or delivered in the drinking water. For complete control of disease, an effective vaccination programme is coupled with an 'all in, all out' policy, where a site is completely cleared of birds before a new flock is brought in, allowing houses and equipment to be thoroughly disinfected. This approach results in a dramatic fall in mortality. In the UK in 1970, about 12% of hens would have died in a well-run laying flock between point of lay at 20 weeks and the end of the laying cycle at 72 weeks. The comparable figure in 1990 in well-managed units was as low as 2–3%. Mortality in US flocks averages about 6.5% by the end of a 60-week laying cycle (National Animal Health Monitoring System, 2000), very similar to the mortality seen prior to initial intensification of production.

Recent trends highlight newly emerging problems in disease control that will affect the industry in the future. These include the increasing resistance of many disease organisms to antibiotics, coupled with regulatory restrictions on antibiotic use, both prophylactic and therapeutic, to preserve their medical value. Vaccination regimes are becoming less effective as new virus variants emerge against which existing attenuated strains offer no protection (Davison, 2003). Even the standard 'all in, all out' policy is becoming less straightforward because many of the most effective biocides are being phased out due to the environmental damage they can cause.

11.6 Control of Photoperiod

The ability to control the photoperiod was an important development for increasing the productivity of poultry. Laying flocks remained dependent on natural light in the UK and many parts of the USA until about 1945, when electricity became widely available on farms. This allowed day length during the winter to be extended by artificial lighting and rate of lay to be maintained at a high level throughout the year. It was found subsequently that rearing on a short photoperiod (such as 8 h light:16 h dark) and then, at point of lay, increasing day length steadily and gradually over an extended period up to about 17 h light:7 h dark gave the best results in maximizing egg size and number. In addition, to be able to provide eggs throughout the year in the numbers required, it became necessary to bring pullets into lay in all months of the year. These two developments demanded a light-controlled environment, with the birds completely isolated from daylight.

Especially in warmer climates, broiler chickens are often still raised in partially open houses that allow less photoperiodic control than in laying houses. However, natural day length is now usually supplemented by artificial lighting to increase the photoperiod to 23 h light:1 h dark, since it was found that this stimulates feeding behaviour and hence growth. However, long light periods are associated with an increase in growth-related problems such as leg weakness, and also may have negative effects on the birds' eye condition (Scientific Committee on Animal Health and Animal Welfare, 2000). For this reason, as well as to reduce electricity costs and improve feed conversion efficiency (Chapter 10), some producers are beginning to use intermittent lighting schedules, with alternating periods of light and dark throughout the day, to slow growth. The amount of light can then be increased to 22–24 h per day during the last week before the birds are processed, which stimulates compensatory growth. The use of intermittent lighting schedules also decreases the incidence of ascites (an accumulation of fluid in the body cavity) and sudden death syndrome, problems that are more common in colder climates where broilers grow more rapidly.

11.7 Control of the Bird's Immediate Environment

There are reasons other than photoperiodic control for intensive housing. In many parts of the world, it can be sufficiently cold in winter to cause problems in outdoor systems, with water freezing and hens with frostbitten combs. Losses of birds to various types of predators can occur, and exposure to disease and parasites is also often greater than in indoor systems.

Food costs make up about 50–60% of the cost of production (Bell, 2002), and keeping and feeding birds indoors greatly reduces food losses to wild birds and rodents. In addition, houses usually provide a warmer environment, which markedly reduces food intake, resulting in increased efficiency of food utilization and thus a reduced cost of production, especially in countries where cold weather is common. In the European Community, for example, average feed costs are US$0.24 and US$0.33 per dozen for eggs from caged and barn hens, respectively, but US$0.40 per dozen for eggs from free-range hens (Fisher and Bowles, 2002).

11.8 Automation

Production costs have also been decreased by a progressive reduction in labour requirements through automating most of the basic procedures in all sectors of the industry. For example, provision of food and water is by automatic feeding and drinking systems. In battery cage systems, manure is removed by scrapers or belts, or falls into a deep pit where it accumulates over the laying cycle. Eggs are collected by rolling on to a belt which conveys them to the end of the house and often on to cross conveyors to an egg processing machine. Hatchery and processing operations are also highly automated, as discussed in Sections 11.10 and 11.14.

Domestic fowls require substantial volumes of water; the daily consumption of a typical layer strain increases from 40 ml at 1 day old, to 120 ml at 6 weeks of age and to 180 ml or more once laying begins (Sainsbury, 1971). In battery cage and broiler production systems, water is generally supplied via either nipple drinkers or drinking cups (Fig. 11.1). In floor systems for hens and breeding flocks, drinkers are traditionally of the bell type, so called because of their shape (Fig. 11.2). One drinker can supply about 100 birds, and as water is consumed the drinker automatically refills, with the disadvantage that, if movement causes spillage or if tilting occurs, the water keeps flowing. This is a potential cause of wet litter, one of the environmental factors with adverse consequences for welfare (Chapter 10). For this reason, nipple or cup drinkers are now more commonly used in broiler production.

Fig. 11.1. Drinking nipple and cup located at the boundary between two cages so that it can be reached by birds from either cage.

Automated feeding systems are usually either chain feeders or, on deep litter, automatically filled hoppers supplying pan feeders (Fig. 11.3). Such systems are often designed to present food in a shallow layer in the bottom of a fairly deep trough. This offers advantages to producers because it reduces production costs by cutting food wastage, but may cause feeding difficulties for birds that have been severely beak trimmed or that have any other form of beak deformity. Automatically filled pan feeders are commonly used for broilers and turkeys.

Since modified environment houses are ventilated entirely or primarily by automatically controlled fans, provisions for back-up power or other mechanisms to prevent ventilation failure during emergencies are critical to prevent flock mortality. However, as automation increases, so too does the scope for controlling the environment. There is now equipment available for automatic regular monitoring of lighting, food delivery, water use, air quality, ventilation rate and other important aspects of the birds' environment, as well as for automatic weighing of broiler chickens (Fig. 9.4).

11.9 Scope and Structure of the Modern Poultry Industries

A number of different types of birds are used commercially for the production of meat, eggs, feathers, hides or other products, including chickens, turkeys, gamebirds (quail, pheasants, and guinea fowl), pigeons, ducks and geese, and emu and ostriches

Fig. 11.2. Bell-type drinkers often used in floor systems. These are readily cleaned and remain automatically topped up, but water is easily spilt should they be tipped.

(Chapter 1). The largest component of poultry production worldwide is undoubtedly that dedicated to production of chicken meat and eggs. Chicken meat accounts for about 20% of world meat consumption, and nearly 50 million tonnes of chicken meat, and another 50 million tonnes of chicken eggs, are now produced annually (Aho, 2002a), mainly in the USA, China and the EU (Food and Agriculture Organization, 1997). In terms of the numbers of birds produced, duck production is the next largest component, concentrated mainly in China and other parts of Asia, followed by turkey production primarily in the USA, the former Soviet Union and France (Food and Agriculture Organization, 1997). Ratite production is the newest component of the industry – domestication and selection of the ostrich did not begin until the latter part of the 19th century, and the industry can be considered to be in the developmental stage, with leather and feathers still the primary products although the market for ostrich meat is increasing (Deeming, 1999a).

In all developed countries, there has been a progressive reduction in the number of producers and an increase in the size of units. Some of these have become large multi-national companies, often with a vertically integrated structure. Their activities range from the growing of grains and feedstuff formulation, through the breeding of specialized strains, the rearing of chicks, housing of laying flocks and production of broilers, to the processing and marketing of eggs and finished meat products.

Fig. 11.3. Automated feeding systems. (A) Chain feeder (and bell drinkers). (B) Automatically filled food hopper.

In some countries, for example Norway and Switzerland, there is a limit on the number of laying hens that can be kept on each farm. In most countries, however, the number of birds (layers or broilers) per farm has progressively increased (see Section 11.5), while at the same time the number of production companies has decreased. In the UK in 1990, for example, it was estimated that about 80% of the UK laying flock was in the hands of only about 300 production companies, while most of the broilers were in the hands of only about 12 integrated companies that supplied the poultry meat market from about 400 growing farms. In the USA, there used to be hundreds of integrators producing broilers, but there are now only about 50, and the top five companies account for 50% of all production (Aho, 2002b). There are fewer than 1000 egg-producing companies in the USA, and around 60 of these companies own 78% of the hens; modern hen complexes can contain as many as 4 million birds (Bell, 2002).

These developments have allowed large integrated production companies to benefit from economies of scale, and the result has been that consumer prices for eggs and poultry meat have decreased. A pound of chicken meat, for example, can now be produced for 88% less (in inflation-adjusted dollars) in the USA than it was in 1944 (Aho, 2002b). However, there may also, correspondingly, have been a tendency at management level for the birds to be regarded primarily as production units, rather than as sentient creatures with their own set of needs.

The facilities and expertise required for the different aspects of poultry production are now generally so specialized that the industry has become separated into distinct sectors; we describe these in the remainder of the chapter.

11.10 Hatchery

Because of the segmentation of the industry, commercial chicken hatcheries usually hatch only either broilers or layers, while only poults are hatched in turkey hatcheries. The hatchery may be owned and operated by the integrator for in-company use only, owned by a primary breeder who sells directly on the open market, or be owned and operated by an independent operator who purchases eggs on the open market or who has a franchise with a major primary breeder.

Perhaps no sector of the industry is more automated than the hatchery sector (Fig. 11.4). In a large automated hatchery, eggs are delivered from the breeder farm in plastic incubator trays or 'flats'. They are mechanically washed and sanitized, automatically incubated, and transferred pneumatically at the appropriate time from the incubator flats into hatchery trays. Vaccinations are mechanically delivered through the shell, and the eggs are passed through a candling machine to assist workers in determining if the eggs are fertile. At hatch, the contents of the hatching trays are tipped out and the chicks are conveyed through a separator machine, which separates the chicks from the shells. The chicks then move through a series of ascending speed belt conveyors, where they are counted with an electric eye and dropped into a chick box, for transport to the growing facility. Chicks may also be mechanically spray vaccinated at this time. The only time that chicks need be handled by humans at the hatchery is if the chicks are to be sexed, needle-vaccinated or beak-trimmed (Chapter 5), or if unthrifty chicks are culled. These procedures are generally carried out at different stations in the hatchery as the chicks move by on the conveyor belt system.

One significant welfare concern in hatcheries is the method of disposal of culled, unwanted or incompletely hatched chicks. In the USA, for example, nearly 260 million chicks are killed in hatcheries each year. The majority of these chicks are killed using a vacuum system that involves sucking the chicks at high speed through a series of pipes to a 'kill' plate, that may or may not be electrified, or to a waste tank (Metheringham, 2000). In some other countries, chicks are killed by

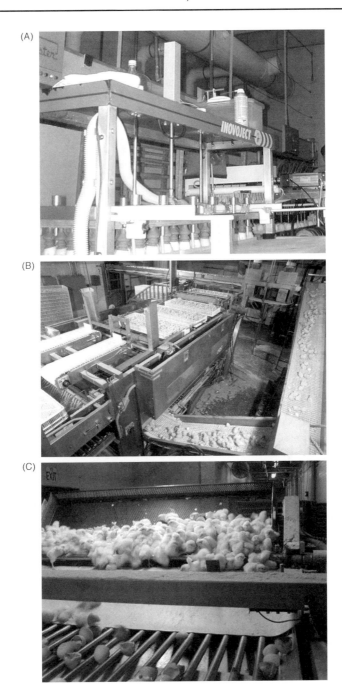

Fig. 11.4. A typical modern commercial hatchery. (A) Many vaccinations are given by injection directly into the egg. (B) When the chicks are hatched, they are carried through a conveyor belt system from the hatching trays to the containers to which they are transported to the grow-out house. (C) After being dumped from the hatching trays, the chicks are separated from their shells by a separator machine (photographs courtesy of Joseph Mauldin).

suffocation or crushing. More humane methods, now widely implemented in the UK and recommended as part of the UK Code of Practice (Department of Environment, Food and Rural Affairs, 2002b), include gas killing (carbon dioxide, argon or a mixture of the two) or instantaneous mechanical destruction using a purpose-designed macerator.

11.11 Egg Production

The egg production sector consists of rearing and laying units. Chicks are hatched and housed either in rearing cages or on deep litter. The system can be single- or multi-stage. In a single-stage system, chicks are brooded, either in cages or on the floor, and kept in the same brood-grow houses until they reach point of lay at 16–18 weeks. In a multi-stage system, once the chicks no longer require the artificial heat of the brooder, they are moved from the brooder houses to grower houses. At point of lay with both systems, the pullets are transferred to layer houses, either within an integrated unit or by being sold on from specialist rearing farms.

Each system has its advantages and disadvantages. A single-stage system, where the birds remain in the same house to point of lay, is believed to be less stressful and cause fewer checks to growth than when they are moved from one cage or pen to another (Sainsbury, 1971). However, it is much more difficult to provide accommodation that can meet the birds' requirements for suitable flooring, appropriate environmental temperatures, and ready access to food and water from day-old to maturity than it is in a multi-stage system.

The majority of hens, 91% in EU countries overall and 99% in the USA, are housed at point of lay in battery cages in groups of 4–10 birds (Fisher and Bowles, 2002). A small, but increasing, proportion (~15% in the UK) is kept in non-cage systems – either 'barn' systems such as deep litter, strawyards, percheries or aviaries, or free-range systems. The effects of these different housing systems, and their associated management, on hen welfare are discussed elsewhere (Chapters 10 and 12).

Most flocks are killed at 72 weeks of age because there is a gradual decrease in egg production and quality with time. As hens age, on average they lay fewer but larger eggs. By about 72 weeks of age, egg production has decreased to 60 or 70% (Fig. 11.5), although this is compensated for by an increase in egg size, so that total egg mass declines only slightly. However, there are also problems of egg quality as the hens age. Egg shell thickness and strength decline, thus increasing the proportion of cracked and broken eggs. In addition, internal egg quality becomes poorer: the proportion of water in the albumen increases with age, resulting in unacceptably watery whites. In some cases, however, the flock is moulted, briefly rested and brought back into production, which results in increased egg number, shell thickness and internal quality. Moulting has been criticized on welfare grounds because it typically involves the withdrawal of feed, sometimes for many days (Section 8.5).

Besides chickens, the only species widely kept for egg production is the Japanese quail. They are also usually kept in cages but the group size is usually larger, between 60 and 80 birds per cage with a stocking density of 120–160 birds/m^2 (Gerken and Mills, 1993).

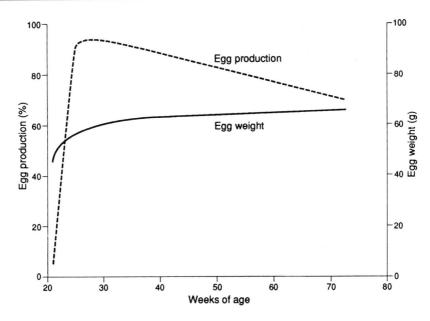

Fig. 11.5. Target egg production values for a medium hybrid laying strain, showing hen-day production and mean egg weight from 20 to 72 weeks of age.

11.12 Meat Production

The broiler industry began in the USA and reached the UK, in the form of both imported birds and management techniques, in the early 1950s. Special strains and hybrids had been developed in the USA for meat production, selected for both rapid growth rate and an efficient food:gain ratio. They were derived primarily from two breeds: White Plymouth Rock and Cornish Game. Growth rate is so rapid that broiler chickens can now grow from a day-old weight of 45 g to a body weight of 2200 g by 42–45 days of age (Fig. 11.6). They consume only 2 kg of feed for every kg of body weight (although note that body weight includes water), a doubling of both gain and feed efficiency since the 1940s (Aho, 2002b).

Almost all broiler birds are kept on littered floors in very large houses (Section 12.4). Many broiler sites consist of several such houses grouped together, e.g. five 20,000-bird houses per site. Flocks are usually reared as mixed-sex groups, but it is becoming more common for the birds to be sexed at 1 day old and divided into separate flocks of males and females. The cockerels require a higher-protein diet, and grow faster and more efficiently for a longer period than the pullets. The pullets thus reach the stage of diminishing returns earlier and are therefore processed and marketed at lower body weights than the males. This strategy is also a response to the demands of the current markets for poultry meat – fast-food companies generally require a smaller bird of around 1700 g, while most birds sold in supermarkets are now sold further processed into parts rather than as whole carcasses, and a larger bird of around 2500 g, with large breast, is therefore

Fig. 11.6. Modern broiler strains are, with turkeys, proportionately the fastest growing of all agricultural animals, from about 45 g at 1 day old to 2200 g by 45 days of age, a 50-fold increase.

desirable. Some birds are also processed when they are still small and sold as Cornish Game hens, while some are grown to a larger size and sold as whole birds called roasters or capons (which in some countries are castrated males).

Chicks are generally brooded in large groups under brooders, with a temporary barrier to prevent them from wandering too far from the brooder. The chickens are kept at very low light intensities, generally less than 5 lux, mainly to reduce locomotor activity, thus decreasing the possibility of the birds piling up and also contributing to increasing the efficiency of feed utilization. However, by decreasing activity, these low light intensities probably also contribute to the development of skeletal disorders, in particular leg problems (Scientific Committee on Animal Health and Welfare, 2000). Stocking densities vary from one country to another, and also depend on the level of environmental control in the house and local climatic conditions. In the USA, stocking densities of around 28–30 kg of bird/m^2 are common, while in Europe typical densities range from 22 to 45 kg/m^2; higher stocking densities can be better managed in fully environmentally controlled buildings. Unless environmental variables are very carefully controlled, high stocking densities can contribute to leg disorders and other health problems (Scientific Committee on Animal Health and Welfare, 2000).

Various litter materials are used for broilers depending on local availability and cost, although softwood shavings are the most common material. To reduce material, labour and litter disposal costs, efforts have been made to develop suitable non-litter surfaces such as wire mesh or slats, or to house broilers in cages, but with limited success. Since growing broilers spend much of their time resting on their keels, they develop breast blisters from the localized pressure due to the cage floor or non-litter surface, which results in welfare problems and carcass downgrading. Leg problems also tend to be worse in cages because of the restriction of movement.

Cushioned floors, where a rubber or plastic overlay covers the wire mesh, can reduce the problem of breast blisters. In some Eastern European countries and in the former Soviet Union, caged broiler production is common.

Turkeys used to be kept mainly on free range, but are nowadays usually kept under very similar conditions to broilers, except that turkey poults are sometimes reared in tier brooders, then moved to littered pens at 3 weeks of age. Special attention must be paid to nutrition: poults require a very high-protein diet in the early stages and can be slow to begin feeding. This problem can be reduced by positioning feeders under bright light.

Ducks are also housed on deep litter under conditions very similar to those of broilers and turkeys, but usually with a slatted area next to the water troughs so that any excess spillage can drain away. Ducks may be housed in cages, or in small groups in pens, if they are to be force-fed for foie gras production. Cannibalism can be a significant problem in ducks, especially Muscovy ducks, and is often dealt with by bill trimming. This is permitted in the UK on the same basis as for fowls, i.e. it should be carried out only when it is clear that more suffering would be caused in the flock if it were not done (Department of Environment, Food and Rural Affairs, 2002b).

Quail grown for meat are also housed under conditions similar to those used for broilers, although cage-rearing systems, particularly for late-stage growing, are under development (Shanaway, 1994). Approximately 70–100 quail are housed per m^2 on the floor in mixed-sex groups, although Bobwhite quail, which are larger than Japanese quail, are generally given more room, particularly if they require flight practice because they are to be released for hunting (Skewes and Wilson, 1990).

Ostrich rearing is more varied than most other types of commercial poultry production (Verwoerd *et al.*, 1999). Chicks may be either artificially brooded or reared by foster parents. Rearing systems range from intensive ones, where small groups of birds are kept in bedded concrete floored enclosures, to more extensive rearing on pasture or in small fenced enclosures. Older ostriches may be placed in feedlots, where they are fed a grower ration. Ostriches' feathers are selectively plucked or trimmed at around 5 months of age to promote uniform feather growth for feather harvesting at slaughter, a procedure that also improves skin quality for the production of leather, but plucking has potentially negative implications for welfare.

11.13 Breeding Sector

The breeding of poultry has become an industry in its own right. Genetic selection has made a major contribution to the success and development of modern poultry production and, because many of the economically important traits are governed in a complex fashion by a combination of genes, the industry has made considerable use of the science of quantitative genetics. Breeding thus includes the development of new strains selected for a wide variety of characteristics. Egg-laying lines have been selected for traits such as egg number, egg size, shell strength, shell colour and low mortality. Meat lines have been selected for growth rate, meat yield, ratio of white to dark meat and rapid feathering. Recently there has been increased emphasis in both cases on selection for efficiency (high output:low feed intake). These specially selected lines are then maintained by the breeder as grandparent stock. Because specialized sire and dam lines are unsuitable for general production purposes, they are crossed with each other to generate heterosis, or hybrid vigour. Commercial hybrids are thus often four-way crosses, being the product of parents that are themselves crosses between two inbred lines of grandparent stock (Fig. 11.7). The eventual output of a breeder is therefore the supply of birds to the meat or egg production sectors.

Considerable effort has been expended on attempts to intensify the chicken breeding industry, but with less success than in the other sectors. Most of the broiler selection lines are still kept in small groups in floor pens, with natural mating. The parent stock is called broiler breeders and is kept to yield eggs that will be hatched to provide the growing broiler chickens. They are housed in larger flocks on deep litter, on farms that are under the management of the broiler integrator. There is one male per 9–10 females, and eggs are laid in nest boxes from which they can be collected either mechanically or by hand. Since broilers have been selected for rapid growth, the body weight of broiler breeders, especially the males, must be controlled to ensure that they do not become obese and thus show reduced fertility (Mench, 2002). Males and females also have different nutritional requirements, so arrangements are usually made to feed males and females separately, by locating female feeders on an elevated slatted area located above the litter floor near the nest boxes, by placing a grid on the females' feeder that does not allow enough room for the male to place his head in the feeder, and sometimes also by rearing the males and females separately until they are sexually mature.

Only in the case of turkeys has artificial insemination become the standard method for breeding, and it was adopted by the industry because selection for breast meat output resulted in such major changes in body conformation that natural mating became extremely difficult for the male. Breeder turkeys are housed in flocks similar to those of broiler breeders, except that the males are kept separate from the female flock, in all-male stud buildings or even on separate stud farms, where they are milked periodically for semen. Lighting, housing density and nutrition can thus be tailored to maximize both male and female reproductive performance.

The focus in the breeding sector has been primarily on manipulating traits affecting growth, reproduction and product quality, rather than traits needed to adapt the birds to intensive production environments. Genetic selection could

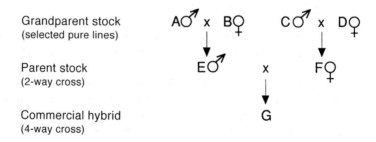

Fig. 11.7. Commercial hybrids are often four-way crosses derived from parent and grandparent stocks.

potentially be used to decrease certain welfare problems in these environments, including problems with social behaviour, fearfulness, nesting behaviour and mating behaviour (Kjaer and Mench, 2003), although there is little emphasis on this at present. There has been some attention paid by the primary breeders recently to decreasing leg problems in broiler chickens by selecting for leg soundness and walking ability, although this seems mainly to have decreased the incidence of one particular kind of leg problem, tibial dyschondroplasia, a heritable disorder involving inadequate ossification of the bone growth plate. Selection of laying-strain chickens for reduced feather pecking and cannibalism (Chapter 5) also appears promising, and may eliminate the need for beak trimming of commercial laying flocks if given sufficient emphasis (Muir and Craig, 1998).

11.14 Processing Sector

Like the other sectors of poultry production, processing has become highly automated, driven by the pressure to process large numbers of birds at high speed. Once a bird is hung on the processing line, killing and preparation of the carcass for shipment to the retail outlet is almost entirely automated, with the bird conveyed past machines that cut the neck veins, scald, de-feather and eviscerate the carcass, and then chill the carcass for packing.

Processing involves a number of stages, the first of which involves loading the birds on to trucks. Prior to loading, the birds are deprived of feed and water for several hours to minimize the potential for contamination of the carcasses with faeces. Chickens, turkeys and Muscovy ducks are usually hand-caught by crews of catchers, and loaded into transport trays or coops, while Pekin ducks and ostriches are herded into the transport vehicle. Transport times are generally relatively short, at least in integrated operations.

Upon arrival at the processing facility, birds may be held in lairage for several hours prior to processing. Maintaining good environmental conditions during lairage is important to prevent mortality, especially mortality due to heat stress. For processing, chickens, turkeys and ducks are usually hung upside down on shackles, and then stunned electrically in a water bath, after which the neck veins are cut. Ostriches are stunned using either a head-only electrical stunner, or with a captive-bolt pistol similar to that used for stunning mammalian livestock. They are then hoisted by both legs and their neck veins are cut.

Catching, transport, holding, loading and unloading, and shackling are undoubtedly stressful procedures for the birds. The number of chickens dead by the time they are received at the processing plant is estimated to range from 0.06 to 3% (Nicol and Weeks, 2000), with 0.25–0.5% typical in the USA (Wabeck, 2002). While this percentage seems small, in absolute terms it is a large number of birds – taking the higher figure, for example, would mean that 40 million broiler chickens die during transit each year in the USA. As many as 25% of the live birds arrive at the processing plant with bruises on their wings, necks or legs (Farsaie *et al.*, 1983). Injuries and trauma are most likely to occur during catching and loading, while thermal stress causing mortality is most likely to occur during transit and lairage (Mitchell and Kettlewell, 1998).

Recently developed equipment that enables the mechanized harvesting of broilers (Fig. 9.3) can offer advantages over typical hand catching in terms of reduced stress and injury, with improved welfare and less carcass downgrading (Lacy and Czarik, 1998). However, harvesters are bulky and can only be used in houses with wide, unobstructed spans and in which the litter depth is not so great that the machine sinks into the litter. Mechanical harvesting is not always superior to human catching, and injury rates are highly dependent on the standard of manual catching and how gently birds are loaded into the transport crates (Ekstrand, 1998).

Improved transport trucks are also becoming available that allow better control of the thermal environment during transit (Fig. 11.8). These trucks incorporate forced ventilation as well as temperature and humidity sensors in critical areas of the load that can alert the driver to conditions that are likely to result in heat stress, thereby allowing corrective action to be taken.

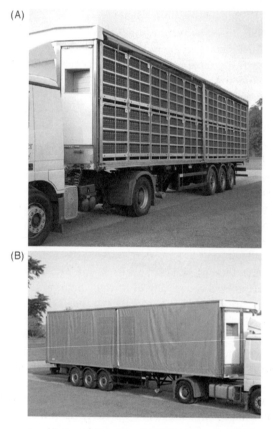

Fig. 11.8. New transport trucks are being used that allow better thermal control during transit. This is the Concept 2000 truck, developed in the UK. (A) Crates are loaded into the truck, and (B) the truck is then operated with the curtains closed. To control ventilation, air enters through perforated mesh sections and is extracted through fans. This reduces heat stress during transit (Kettlewell and Mitchell, 2001; photographs courtesy of Malcolm Mitchell and Peter Kettlewell).

Concerns have also been raised recently about the use of electrical stunning in processing plants. Electrical stunning is associated with carcass and meat quality defects, and may not result in all birds being adequately stunned prior to slaughter, particularly if low currents or low voltages are applied (Raj, 1998; Savenjie *et al.*, 2002). Because their wings hang lower than their heads when they are shackled, turkeys may also experience pre-stun shocks. Controlled atmosphere stunning (CAS) systems, in which broiler chickens and turkeys are stunned using carbon dioxide, argon and/or nitrogen (Raj, 1998), have been developed and are now being adopted in some European processing plants. Gas stunning not only improves carcass quality but also allows the birds to be stunned while still in their transport crates so that the birds are not stressed or injured during uncrating and shackling. A disadvantage of this is that it is difficult to identify dead and moribund chickens in the transport crate, and these birds should not be processed because of food safety considerations. For this reason, some CAS systems have been designed such that birds are first tipped out of the crate on to a conveyor so that they can be examined before they enter the stunning tunnel.

End-of-lay hens (sometimes called 'spent hens') are still generally sent to processing plants. However, the availability of inexpensive broiler meat, coupled with what is considered to be the relatively poor quality of meat from these hens, has resulted in a diminished market for hen meat for human or animal food. Hens are now increasingly either killed on-farm or transported to specialty and live-animal markets, where meat from birds at the end of lay is still valued for its taste. In the USA for example, about 9% of hens are killed on-farm and rendered, while 11% are sent to live-animal markets (National Animal Health Monitoring System, 2000). Both of these methods of hen disposal raise welfare concerns. It is difficult to kill large numbers of hens on-farm humanely, although there has been some research on potential methods, including carbon dioxide, electrocution and maceration (Newberry *et al.*, 1999). Hens have osteoporosis due to their high rate of egg laying, which means that their bones are easily fractured or broken during depopulation and transport. Transit times and distances for end-of-lay hens can be long, even when they are transported to processing plants, and this results in high mortality during transit (Newberry *et al.*, 1999).

11.15 Welfare Considerations

Welfare regulations and recommendations are now beginning to direct the industry in terms of stocking densities, banning of practices such as beak and toe trimming and the use of feed withdrawal to induce moulting, setting atmospheric standards for dust and ammonia, providing safeguards if automated control systems malfunction, and improving bird handling, transport and slaughter (Chapter 13). There is also growing awareness of a requirement for first class management and husbandry. This includes the need for attention to detail, the importance of a recording system (especially one which can provide an early warning of changes in variables such as daily water intake), a feeling for stock and the ability to identify and interpret behavioural changes, and the ability to diagnose and quickly correct both mechanical and biological problems.

12   Systems

12.1 Summary

- Simple farmyard systems are mostly for domestic consumption. Free range originally developed with small, moveable huts, but modern commercial systems involve large flocks in fixed houses, and only a minority of birds go outside.
- Covered strawyards are naturally ventilated, open-fronted buildings where the floor is covered in straw. When used for turkeys, they are often called pole barns. Light is supplemented and stocking density low. Deep litter houses are fully enclosed, with part slatted floors permitting higher stocking densities. Breeding ducks are usually kept in small pens on deep straw, but large groups of fattening ducks are housed on slats or wire.
- Multi-bird (3–8) cages for laying hens have many economic advantages: low labour requirements, low food intake and generally clean eggs. Management is straightforward and working conditions are good. There are some welfare advantages, but also major disadvantages – primarily severe behavioural restriction. Over the last decade, there have been improvements in cage design: feed troughs, drinkers, floors, fronts and partitions.
- Furnished cages provide most of the advantages of conventional cages while removing many of the behavioural restrictions. They contain perch, nest box and littered area, and provide more height and area per bird.
- Multi-level systems increase the effective vertical space of the house by allowing birds to use different levels above the floor. Aviaries have tiers of slats or mesh, percheries have perches at a range of heights, and the tiered wire floor system resembles a cage house with the partitions removed. Such systems have been in use in Switzerland since cages were banned in 1992, and most of the initial problems have been overcome.
- In this area, economics and welfare overlap: producers have to conform to regulatory constraints in order to qualify for particular categories of system,

which in turn means their eggs will command higher prices. The EU is likely, in the near future, to recognize only three categories: free range, perchery (or barn) and cage eggs.

12.2 Farmyard

Small-scale operations in which hens, ducks or geese wander freely during the day are widespread on farms and rural properties. This is the only truly free-range system, in that the birds are completely unrestricted in their movements, except that they are usually shut up at night for protection from predators. The housing needs some bedding and may have roosts. With hens or ducks kept for eggs, nest boxes are also provided. Shutting birds in and letting them out takes some time; although they may forage widely, another daily task is providing supplementary food. These jobs may be labour intensive depending on the number of houses and birds, but the system falls outside normal economic analysis, since such labour is not costed and products are usually only for domestic use. Many of the principles and problems found in farmyard flocks recur in larger systems. For example, design and management of nest boxes for farm poultry, and remedies for birds failing to use them, have been considered over many decades (Chapter 6). They continue to be important in all laying systems except battery cages. Even with small flocks, damage to ground is likely unless the house opens on to slats, concrete or gravel, and larger flocks cause more damage. With the development of commercial poultry keeping late in the 19th century, there were two main developments: pasture-based systems, called free range, and house-based systems, called semi-intensive (Hewson, 1986).

As already mentioned (Section 10.3), farmyard poultry systems have some similarities to the wild – to the environment of evolutionary adaptation (Section 3.3). They necessarily take the behaviour of the birds into account. Few subsequent developments in system design have taken poultry behaviour as a major premise. Rather, they have usually adapted previous systems in attempts to use currently available technology to improve economic performance. There have been two exceptions to this trend. Work on furnished cages for laying hens, first on the get-away cage and then on other forms, was intended to ameliorate the behavioural restriction imposed by conventional cages (see Section 12.6). Secondly, aviaries and percheries (see Section 12.7) put into practice the suggestion of McBride (1970) that as wild fowl live in three-dimensional conditions, housing could be designed in a similar way. We concluded in Chapter 10 that a challenge exists, to design production systems in which birds acting for their own advantage will also be acting for the advantage of the producers. It will become apparent in this chapter that this challenge has only partially and intermittently been met.

12.3 Free Range

In early forms of free range, damage to ground and build-up of disease were avoided by using small, moveable houses. These were still primarily for use at night, although additional light could be given inside to extend natural day length. Supplementary food was given in hoppers outside and the houses were shifted every few days or so to fresh ground. Alternatively, birds were kept in 'fold units' – combined houses and pens that were also mobile. Highly labour intensive, these approaches were adapted by incorporating fixed housing (Fig. 12.1) big enough for birds to be fed inside. They also obtain some nutrition from the outdoor area, particularly on pasture. However, a similar arrangement without vegetation is being adopted in some conditions that cannot provide it, for example in organic egg production in some parts of the USA. In any case, consumption of provided food is actually higher on range than in housing, at least in temperate countries (Hughes and Dun, 1986), because of increased activity and lower temperature.

One problem with free range, with any species of poultry, is that the birds use the area near the house heavily but the rest much less. Use of slats or other arrangements near the entrances to the house may reduce ground damage, but low stocking density may still be necessary to prevent the development of unpleasant conditions. Good drainage of the ground away from the house is also important, together with use of several areas in rotation. One possible arrangement is to have a house surrounded by several such areas, with pop-holes for the birds next to each. Different pop-holes can then be opened as appropriate. Areas not in current use may be left fallow to recover or can be grazed by other animals, which helps to maintain a good turf. If one area is used permanently, stocking density must be low enough to prevent not only destruction of vegetation but also build-up of disease. For laying hens, some advisers recommend a maximum of 300 birds/ha (Hann, 1980) and some up to 400 birds/ha (Elson, 1985). However, they are probably considering relatively small flocks. EU trading standards permit eggs to be sold as free range from flocks with up to 1000 birds/ha, while specifying that the ground

Fig. 12.1. Free range with fixed housing (photograph courtesy of Arnold Elson).

must be mainly covered with vegetation (Table 12.1). This is achievable because, in the large flocks now common commercially, only a minority of birds actually go outside. This is partly because cover is rarely provided, despite the fact that chickens evolved in forests and are cautious because of their vulnerability to aerial predators. Even with low use of pasture, the establishment of hard-wearing grasses is important.

For eggs to be described as free range, EU regulations also state that housing must have the size and facilities appropriate to deep litter or percheries (Table 12.1), while in the USA there are no detailed specifications. Increased costs compared with cages (Chapter 14) may be offset by increased income from the premium available on free-range eggs. In some countries, this is the only system for which such a premium is available, since consumers apparently find other categories, such as those in Table 12.1, confusing or unattractive. A proportion of consumers are prepared to pay more for free-range eggs, but the reasons for this are not always clear. One reason is probably that hens on free range are presumed to have better welfare. At low stocking density and with good management, this is probably a reasonable presumption, but these conditions are not always met. There is also a potential exception, in that on range as in all non-cage systems there is a risk of cannibalism and so birds are usually beak trimmed. In cages, cannibalism is rare and beak trimming is less necessary; however, it is still often practised. Other reasons

Table 12.1. Criteria defined by EU trading standards regulations (Commission of the European Communities, 1985) for labelling of eggs.

Label	Criteria
(a) Free range	Hens have continuous daytime access to open-air runs
	The ground to which hens have access is mainly covered with vegetation
	The maximum stocking density is not greater than 1000 hens/ha of ground available to the hens or one hen/10 m^2
	The interior of the building must satisfy the conditions specified in (c) or (d)
(b) Semi-intensive	Hens have continuous daytime access to open-air runs
	The ground to which hens have access is mainly covered with vegetation
	The maximum stocking density is not greater than 4000 hens/ha of ground available to the hens or one hen/2.5 m^2
	The interior of the building must satisfy the conditions specified in (c) or (d)
(c) Deep litter	The maximum stocking density is not greater than 7 hens/m^2 of floor space available to the hens
	At least a third of this floor area is covered with a litter material such as straw, wood shavings, sand or turf
	A sufficiently large part of the floor area available to the hens is used for the collection of bird droppings
(d) Perchery (barn)	The maximum stocking density is not greater than 25 hens/m^2 of floor space in that part of the building available to the hens
	The interior of the building is fitted with perches of a length sufficient to ensure at least 15 cm of perch space for each hen

people prefer free-range eggs are less well founded: perceptions that they taste better, are better nutritionally and are more likely to be free of disease such as *Salmonella* than eggs from cages. If anything, eggs are more likely to be infected on range than in cages.

In some countries such as France, free-range broilers and turkeys are common. This is for reasons of taste, which includes meat texture, rather than welfare. Texture is influenced by exercise, which considerably affects muscle development, and by rate of growth: slow-growing strains are often used. In this case, a premium for the free-range product may have more justification. As with laying hens, use of available pasture may be very uneven. For example, some flocks of turkeys on range in the USA have access to several hectares, but severe local crowding may occur.

EU trading standards include another category called semi-intensive, with up to 4000 birds/ha (Table 12.1). This system probably survives mainly for small backyard flocks, where it is more convenient to set aside a small amount of land for their exclusive use than to have them roaming at large.

Ostriches and other ratites are mostly kept in various forms of free range: breeding pairs and triplets in enclosures on pasture, with housing for inclement weather, and growing birds in enclosures or feedlots.

12.4 Floor Housing

Strawyards were introduced for laying hens as an intermediate between primarily outdoor and wholly indoor systems. Among other advantages, they avoided the need to shut up birds at night. They most often used an existing farm building, usually open-fronted, with an adjoining yard. The yard and part or all of the house had a thick bedding of straw as litter, with wire or slats in the rest of the house. Capital costs were low, but wet straw in open yards resulted in two main problems: build-up of disease and high proportions of dirty eggs. Fully covered strawyards, also called covered yards, are now the rule (Fig. 12.2). Similar houses are also used for turkeys, usually called pole barns. These may still be converted from existing buildings, or they may be purpose-built. They retain the same principle of using natural ventilation, with the upper part of the front of the house, or of several walls, being open. Solid lower parts to the walls, though, reduce draughts at bird height and screens can be used over the openings in bad weather. Insulation is minimal, except sometimes over a roost area for increased protection at night, and the system is only suitable in temperate conditions where extreme temperatures are rare. Because the birds are within a house, though, supplementation of natural light is possible. Flocks generally comprise several hundred birds, e.g. 400–600. Use of straw as bedding tends to limit this system to arable areas where straw is cheaply available, although other litter can also be used in similar houses. Straw may have some nutritional advantage over other litter. In one long-term study, birds in a strawyard ate slightly less than those in cages, with better conversion efficiency (Sainsbury, 1980). This must have been because they obtained some energy from material in the straw, such as grain and insects. In general, however, the increased activity in a yard and the low temperature that occurs for at least part of the time result in increased food consumption (Gibson *et al.*, 1988).

Fig. 12.2. Part of a covered strawyard, with natural lighting and ventilation and a raised roosting platform (photograph courtesy of Arnold Elson).

Under EU trading standards, eggs from strawyards can be sold as deep litter eggs if stocking density is up to 7 hens/m^2 (Table 12.1). In practice, different authorities recommend less than 4 hens/m^2 (Sainsbury, 1980), or no more than 3 hens/m^2 (Ministry of Agriculture, Fisheries and Food, 1987a). A study of densities up to 6 hens/m^2 found increased cannibalism and other problems above 4 hens/m^2 (Gibson *et al.*, 1988).

The system specifically called deep litter is not completely distinct from strawyards, but it is usual for deep litter houses to be more fully enclosed. With automatic ventilation, this allows more precise temperature control. In many cases, natural light is also excluded to allow the use of photoperiods shorter than day length. For this reason, deep litter is often used for the rearing of poultry, even if they are to be housed in a different system later.

In any litter-based system, birds defecate on the litter, and the consequences of this are important. Effects depend to a large extent on the behaviour of the birds, which is an integral feature of the functioning of the system. Thus, the usual reaction of birds to loose litter is to peck and scratch in it. As a result, faeces do not simply accumulate but are dispersed. They may then dry out and be broken down by bacterial action. When this happens and the litter remains dry and friable, it is said to be 'working'. If litter becomes wet, however, because of water spillage, low temperature or high stocking density, it can also pack down and become solid. Either condition inhibits pecking and scratching, so a slight problem can rapidly become worse. Unpleasant conditions, including high ammonia, then develop, and foot damage and disease are likely, so good litter management is very important. The most common litter used is wood shavings, although other bedding materials such as sand, corncobs and groundnut hulls are used increasingly in some parts of the USA.

Stocking density can be higher if part of the floor area is slats or wire mesh, so that the proportion of droppings accumulating in the litter is lower. This is achieved either by having a pit below the slats, or by having a raised, slatted platform (Fig. 12.3), with droppings accumulating on the solid floor. Birds are encouraged to roost on the slats at night, if necessary by moving them there nightly when they are first housed; this necessity can be avoided by rearing them with roosts (Section 5.4). Drinkers placed over the slats reduce the risk of wet litter. With half the house area as slats, laying hens have been stocked at 11 hens/m^2 (Appleby *et al.*, 1988b) or higher, with no litter problems. Similar houses are also used for broiler breeders in North America and layer breeders in Europe. Broilers and European broiler breeders, however, are usually housed entirely on litter. Flock size varies with type of bird. In Europe, breeders are most often housed in groups of 4000–5000 and broilers in groups of 10,000–20,000. In North America, breeder houses usually hold 7000–8000 birds and broiler houses up to 25,000 birds.

Deep litter is also the most common system for housing breeding ducks. Straw is often used as litter, because ducks drink a lot of water and produce wet droppings. These can be taken up by open straw, but frequent renewal of litter is necessary. In contrast to the large flock sizes for hens, ducks are housed in breeding groups in small or medium pens. Growing and breeder turkeys in the USA are also commonly kept in deep litter houses.

Sale of eggs as deep litter eggs in the EU limits stocking density of laying hens to 7 hens/m^2 with at least a third of the floor as litter (Table 12.1). At this density, the system is only economical if a premium can be obtained for the eggs (Chapter 14), while at higher densities there may be problems with restriction of movement or cannibalism (Appleby *et al.*, 1989).

Fig. 12.3. An experimental deep litter house for laying hens. On the left is a slatted platform, with droppings accumulating below; on the right are raised nest boxes.

Brooding and fattening of ducks is usually carried out on slatted or wire floors, because of the problems of maintaining deep litter in good condition. Similar systems have also been tried for laying or breeding hens, but have many behavioural problems including floor laying, cannibalism and hysteria. Furthermore, in the EU, eggs would have to be labelled as coming from a perchery, yet single-tier systems would cost more and compete poorly with multi-tier systems.

12.5 Cages

Cages were introduced originally for single laying hens to allow recording of individual egg production and culling of poor layers. Later, several birds were placed in each cage, and group sizes of from three to eight are now typically used (Fig. 12.4). This results in reasonable economy on capital costs per bird (helped in the USA by tax breaks for equipment purchase), and cages have other economic advantages, such as reduction of labour for maintenance and reduced food intake. The former includes use of automatic methods for feeding, manure removal and egg collecting. The latter is partly caused by increased temperature resulting from high stocking density in the house. Cages also avoid some of the behavioural problems of high-density floor systems, in two ways. First, the cage environment controls certain aspects of behaviour, such as egg laying. Eggs are laid on the sloping floor of the cage and roll out for collection, so there is no need for nest boxes for egg collection and no problem associated with the fact that birds sometimes fail to use nest boxes, as is found in floor systems. However, such behavioural restriction also has many adverse effects (Section 8.6; Appleby and Hughes, 1991). Secondly, social problems associated with large group size, such as aggression and major outbreaks of cannibalism, are reduced. It is apparent that with regard to the welfare of laying hens, cages have both disadvantages and advantages. One major advantage is the separation of birds from their faeces and from litter, thus reducing disease and parasitic infections.

Cages for larger groups of layers, called colony cages, have also been used. Similar cages are used for rearing domestic fowl and for housing small groups for mating if pedigrees are needed. In both cases, cage floors can be flat. Turkeys are not caged commercially, except for some breeding turkey hens, in which artificial insemination is routine.

Working conditions for operatives are often better with cages than with other systems: automation reduces the amount of physical labour, and dust and ammonia are usually less prevalent.

For any stock, food is supplied in a trough in front of the cage and water in an automatic system such as a nipple, cup or trough line through the cages (Fig. 12.5). In old-fashioned cages, still in use in much of the world, cage fronts consist of thin vertical wires that cause feather abrasion during feeding, and other faults in design often cause birds to become trapped and suffer injury or death (Tauson, 1985). Modern cages for laying hens have simplified fronts with horizontal bars (Elson, 1988) and solid cage sides (Tauson, 1989), reducing feather damage. Injuries are also less prevalent (Tauson, 1988). Other improvements in cage design include the use of improved feeding systems to reduce food spillage. Cages are usually arranged

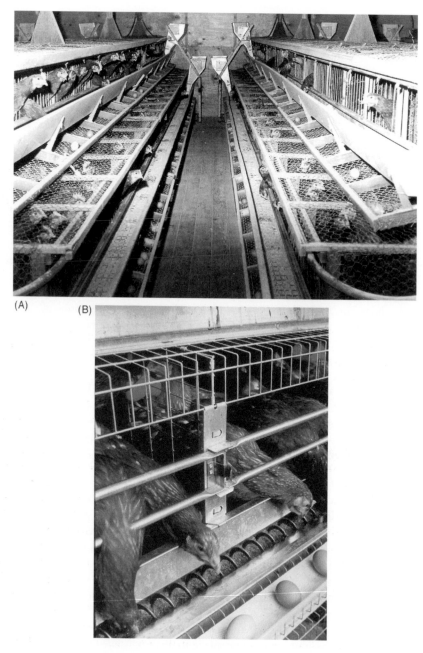

(A)   (B)

Fig. 12.4. Types of cages. (A) An old-fashioned cage house, with manual food distribution and egg collection, vertical-barred fronts and hexagonal wire-mesh floors. The cages are fully stepped over a deep pit; grids in the food troughs reduce wastage. (B) Modern cages for laying hens. Simple doors are easy to operate and reduce feather wear; the chain that delivers food also reduces wastage (photograph courtesy of Arnold Elson).

Fig. 12.5. Section through a laying cage. At upper left is a drinking nipple with a drip-cup; at lower right is the food trough, with a chain at the bottom that is pulled through to deliver food. Eggs roll under the food trough (photograph courtesy of Arnold Elson).

in tiers. These are either vertically stacked, with faeces removed by a belt or a scraped shelf between the tiers, or arranged step-wise so that faeces fall into a pit. The large number of cages in one house necessary to accommodate a flock is called a battery of cages and laying cages are often called battery cages. It is common to have 20,000 birds per house in Europe and 60,000 or more per house in the USA. Space allowances for laying hens in cages vary in different countries from about 300 cm^2 per bird upwards, but in the EU there was a statutory minimum of 450 cm^2 (Commission of the European Communities, 1986) until 2002 and 550 cm^2 from 2003 (Commission of the European Communities, 1999; see also Section 13.4). The former figure was matched in the USA by the McDonalds Corporation, which from 2000 required its egg suppliers to provide this amount of space for their hens, and some other buyers are now following suit. United Egg Producers, the main egg commodity group in the USA, has also produced a set of husbandry guidelines (United Egg Producers, 2002) recommending that their members provide a similar amount of cage space, which the majority of producers have agreed to do voluntarily. The effect of any specific space allowance, however, will partly depend on the number of birds per cage and the actual cage size and configuration and partly on other factors, including the size and activity of the hens concerned. White egg layers are generally smaller than medium hybrids, which lay brown eggs, so that recommended space allowances seem slightly less restrictive. However, light hybrids are more active than medium, especially before laying (Chapter 6), so the restrictions of the cage environment may be more important.

The EU Directive on Battery Hens also specifies the minimum height for cages, of 40 cm over 65% of the minimum area and 35 cm for the remainder. It requires a minimum of 10 cm of feeding space per bird and a maximum floor slope of 8° for rectangular mesh. There are also regulations covering management, e.g. the requirement that all automatic and mechanical equipment should be inspected at least once daily. The legislative emphasis on caged birds reflects both the prevalence of this system (used for >90% of layers) and its nature: birds are closely affected by every feature of the system, and more so than in any other system.

12.6 Furnished Cages

The design of cages for laying hens has changed often to improve economic performance. More recently, there have also been modifications specifically to ameliorate welfare problems. Perches have negligible cost and encourage normal roosting behaviour (Tauson, 1984; Elson, 1985). Depending on design, they may reduce foot problems and bone weakness (Hughes and Appleby, 1989; Duncan *et al.*, 1992). They may also reduce food intake (Braastad, 1990). However, a drawback is that, if this is the only modification made, they tend to increase the number of eggs that are cracked or dirty. Attention has also been paid to reducing overgrowth of claws, by attachment of an abrasive strip to the egg guard behind the food trough (Tauson, 1986).

Early attempts at more radical cage design included the get-away cage (Elson, 1981; Wegner, 1990), which incorporated perches and nest boxes and a greater freedom of movement vertically as well as horizontally. However, with groups of up to about 60 birds, aggression and feather pecking were often severe, and there were hygiene problems because birds sometimes defecated on each other. Subsequent work therefore emphasized retention of the main welfare advantages of conventional cages: small group size and hygienic conditions.

What have come to be called modified, enriched or furnished cages provide increased area and height compared with conventional cages, and also a perch, a nest box and a litter area (Fig. 12.6: Sherwin, 1994; Abrahamsson *et al.*, 1995, 1996; Appleby, 1998). The term 'furnished cages' is probably best, being simply descriptive. Following large-scale adoption of such cages in Sweden, results from commercial flocks are now becoming available (Tauson, 2000; Tauson and Holm, 2001).

Several commercial models of furnished cages are available, but work continues on suitable designs. One large-scale trial found that front rollaway nests lined with artificial turf were successful, attracting settled pre-laying behaviour and up to 93% of eggs. Behaviour was more unrestricted and varied, and physical condition was better, in furnished than in conventional cages. There has also been no cannibalism in studies of furnished cages. However, egg production will cost more, partly because more eggs are downgraded and partly because of capital costs. Litter areas must be fitted with gates that the birds cannot open from outside, to prevent birds from roosting and nesting there, but gates for nest boxes were found to be unnecessary (Appleby *et al.*, 2002).

Fig. 12.6. Furnished cage for laying hens. Doors for the nest box (at lower left) and dust bath (at upper left) are operated by sliding rods in the roof of the cage, which also includes a perch.

In 1996, the Scientific Veterinary Committee of the European Commission listed welfare benefits and deficiencies of cages and non-cage systems and concluded (p. 109) that:

> To retain the advantages of cages and overcome most of the behavioural deficiencies, modified enriched cages are showing good potential in relation to both welfare and production.

On this basis, the Commission passed a Directive (Commission of the European Communities, 1999) requiring that by 2012 all laying cages shall be 'enriched', providing at least the following: 750 cm^2 per hen, of which 600 cm^2 is 45 cm high, a nest, a littered area for scratching and pecking, 15 cm of perch and 12 cm of food trough per hen, and a claw-shortening device. The Committee, now the Scientific Committee for Animal Welfare and Animal Health, will report to the Commission in 2004 on prospects for implementation of the Directive.

12.7 Multi-level Systems

The principle of allowing birds to use different levels, as in get-away cages, was then adapted to larger pens or whole houses. In the UK, aviaries were used first for breeding flocks of fowl, then adapted for layers. The aviary is basically a floor system, but with tiers of slats, wire or plastic mesh to increase the use of vertical space in the house. These tiers must either be strong enough for people to walk on, or be arranged with passages for workers. Ladders between them are intended to encourage free movement of birds. Part of the floor is usually litter, although some trials in Switzerland investigated aviaries without litter (Amgarten and Mettler, 1989; Matter and Oester, 1989). Drinkers are placed over slats and feeders are widely distributed. Nest boxes are placed as accessibly as possible, but floor laying is sometimes a problem (Hill, 1983). There are two other potential problems that have raised doubts as to whether the system is generally beneficial to poultry welfare. First, in common with other floor systems, litter can become wet and this has on occasion led to severe foot damage (Hill, 1986). Secondly, some birds can defecate on others. This may partly account for the intense feather pecking which has occurred in some flocks, to the extent that birds were completely denuded (Hill, 1983). One development in Switzerland, Sweden, The Netherlands and the UK has been the use of manure belts under the upper tiers to prevent this problem.

Various arrangements of tiers have allowed experimental stocking densities of up to 22 birds/m^2 of floor space (Elson, 1985). Within the EU, if the density is more than 7 birds/m^2, eggs must be sold as perchery eggs; otherwise they can be labelled as deep litter eggs (Table 12.1). Group size is usually of the order of 1000 birds.

The perchery, a system for laying hens, uses the vertical space of houses by providing perches rather than platforms, arranged on a frame so that birds can jump up or down from perch to perch (Fig. 12.7). Some experimental houses called aviaries have also used perches (Oester, 1986; Wegner, 1986), so the terms overlap. Similar overlap occurs in other European languages, and some confusion is caused by the fact that EU trading standards use the term perchery in English (Table 12.1), but the equivalent of the term aviary in other languages (e.g. 'voliere' in French; perchery translates as 'perchoir'). Use of perches has allowed stocking densities of up to 18 birds/m^2 experimentally, with good results (Michie and Wilson, 1984; McLean *et al.*, 1986; Alvey, 1989). In these studies, there was ample perch space and litter on part of the floor, which was used intensively at certain times. There were few behavioural problems (McLean *et al.*, 1986). However, the EU requirements for perchery eggs allow up to 25 birds/m^2 and do not include providing litter (Table 12.1). Without litter, birds do not use the floor fully and the minimal requirement of 15 cm of perch space per bird does not provide complete freedom of movement. Commercial firms applying these standards have encountered problems such as cannibalism (Harrison, 1989) and non-laying birds occupying nest boxes. Nevertheless, in some countries, there is a market for eggs labelled as coming from this system, providing a premium for the producer. Early experimental percheries were pens with about 120 birds; subsequent versions contained flocks of over 1000.

The tiered wire floor system developed in The Netherlands (Ehlhardt and Koolstra, 1984) is similar to the aviary in using tiers to increase vertical use of house space, but was designed with the requirement for matching the stocking density of three tiers of cages. The system resembles a cage house with the partitions removed: there are rows of narrow tiers with passages in between the rows and a manure belt under each tier (Fig. 12.8). Nest boxes are provided, but the tiers are sloping in case eggs are laid on them. Perches are mounted over the top tier and food and water are supplied at all other levels except the floor, which is covered with litter. With stocking density up to 20 hens/m^2 of floor area, performance similar to that of caged birds

Fig. 12.7. Perchery for laying hens, with a framework of perches from which birds feed and drink. There is litter on the floor at lower right, and several tiers of nest boxes at rear left (photograph courtesy of Arnold Elson).

has been recorded (Centrum voor Onderzoek en Voorlichting voor der Pluimvee-houderij, 1988), although not consistently from flock to flock.

Fig. 12.8. Tiered wire floor system, viewed from the end. The doors give access to passages between the tiers, for operatives. There is litter on the floor and there are nest boxes on the right (photograph courtesy of Arnold Elson).

Tiered wire floor systems and other variants of aviaries and percheries are almost universal in Switzerland, where conventional battery cages were banned from 1992 (Oester and Fröhlich, 1989). The Dutch design was adapted in a number of ways, and marketed in Switzerland and elsewhere under trade names such as 'Natura', 'Multifloor' and 'Voletage'. Problems such as floor laying, disease and cannibalism were quite widespread initially, especially as beak trimming is not permitted, and the management of stock is more difficult and demanding than in cages (Amgarten and Mettler, 1989). However, it is reported that performance and welfare of the birds, while relatively poor at first, have improved with experience (Fröhlich and Oester, 2001). Not being in the EU, Switzerland can restrict imports of cheaper eggs; imports are needed, however, because the country is not self-sufficient in eggs. It also directly subsidizes its egg farmers.

A subsequent development in Switzerland, also subsidized, has been the combination of these systems with either free range or a 'terrace' along the side of the house with open-mesh walls. With relatively small flocks, free-range birds use the outside area extensively, and terraces are also well used.

Germany decided in 2001 that, in the context of a Europe-wide phasing out of battery cages, it will also disallow furnished cages within its own borders, producing a situation similar to that in Switzerland. At the time of writing, The Netherlands is considering a similar move.

12.8 Marketing and Labelling

Superficially, it seems unsurprising that, as mentioned at the start of this chapter, design of poultry systems has been affected more by economics than by behaviour and welfare. As we have emphasized, however, performance of systems is integrally affected by the behaviour of the birds. Therefore, more attention is now being paid to behaviour, belatedly, as well as to welfare. In one major area, however, consideration of economics and of welfare has overlapped: marketing of eggs in the EU. There are fewer laws in the EU or its constituent countries on how poultry must be kept than is commonly supposed, e.g there is no law on maximum stocking density in floor systems. Yet producers use labels to describe the systems in which eggs were produced, because consumer concerns about welfare of laying hens are prominent. So regulation by the EU of the criteria for those labels has had a major influence on producers' choice and naming of systems and on management – notably stocking density – with profound effects on behaviour and welfare. Thus, producers keeping hens on deep litter tend to limit stocking density to 7 birds/m^2 so that they can market their eggs as coming from deep litter and get a premium for them (Chapter 14). This influence is likely to intensify, as the European Commission currently is considering revising and simplifying the criteria in Table 12.1, with just three categories of free-range, perchery and cage eggs.

13   Politics

13.1 Summary

- Animal welfare legislation depends on information, both scientific and practical, but also both follows and influences public opinion. For example, in the UK, the Prevention of Cruelty Act established cruelty to animals as a criminal offence in 1911. It changed public attitudes, both by deterring abuse and by changing perceptions.
- Attitudes vary considerably between countries, being, for example, more positive towards animal welfare in northern than in southern Europe. However, the national legislation of European countries is being gradually overtaken by that of the EU, which has developed from the 1976 Convention on the Protection of Animals. It has produced Trading Standards Regulations for labelling eggs, a 1986 Directive for laying hens setting out minimum stocking rates and construction requirements, and a 1999 Directive phasing out all cages, unless furnished, by 2012.
- In the USA, most State legislation exempts farm animals from protection if this conflicts with normal industry practices. There are few federal laws governing animal welfare and none apply to poultry; this is surprising given that public opinion is generally supportive of the humane treatment of animals.
- Trade organizations tend, usually after a lag period, to follow public opinion. In the USA, initial opposition to welfare improvements has begun to wane, partly because of pressure from retailers. However, the International Egg Commission is attempting to limit the further spread of cage bans.
- There are numerous animal protection societies in Europe, such as the RSPCA, and in the USA, such as the Humane Society of the United States, as well as international ones, such as the World Society for Protection of Animals. Many have policies aimed at improving conditions for farmed poultry.
- Increasing concern for animal welfare may be partly caused by the move away

from rural living, partly by increasing affluence and partly by the spread of intensive agriculture. It seems likely that this concern will not wane.

13.2 Attitudes and Legislation

Given the complex issues involved in poultry management and welfare, as conveyed by the contents of this volume up to this point, it is not surprising that political and legal decisions affecting how poultry are treated are difficult. First we should consider the relationship of such decisions to public attitudes. On the one hand, the way in which people regard the animals they keep for food, companionship, experimental purposes or sport is reflected in the laws that have been passed to ensure the protection of those animals. On the other hand, there is never unanimity in public attitudes on such subjects, and anyway it would be unworkable for politicians to base all their actions on opinion polls. So as Knierim and Jackson (1997, p. 249) point out:

> Legal action in the field of animal welfare depends to a large degree on public opinion but is also dependent on the input from any experts able to provide solid scientific and practical information submitted on the basis of knowledge of the legal and political context of the measure.

Perhaps it would be better to say that politicians may be influenced by scientific *or* practical information as people perceived as experts may come from various sectors such as the academic community, the poultry industry and the retail sector, and may have special interests in influencing the decisions that are taken. Another reason why legislation does not just reflect public opinion is that sometimes it is appropriate for politicians and other leaders to lead rather than to follow. In any event, legislation affects public attitudes – whether of the minority or of the majority – as well as reflecting them.

These processes can be illustrated by historical developments in the UK. In 1822, the UK Parliament passed a Bill forbidding the ill-treatment of horses and cattle at the instigation of Richard 'Humanity' Martin against the ridicule of many other Members of Parliament. In 1835, Princess Victoria became patron of the newly founded Society for the Prevention of Cruelty to Animals, which thereby became the Royal Society for the Prevention of Cruelty to Animals (RSPCA). A second law was then enacted giving some additional protection to all domestic animals and outlawing bear, badger and bull baiting, and dog- and cockfighting. By 1911, opinion on the welfare of farm animals had developed sufficiently to prompt passage of the Protection of Animals Act. This established cruelty to domestic animals as a criminal offence, evidenced by the infliction of unnecessary suffering, and it remains the basis of UK animal welfare legislation today. An important feature was that intention or otherwise to commit cruelty was immaterial: the only question was whether pain or suffering was inflicted and, if so, whether there was good reason. Six classes of offence were recognized:

1.   To beat, kick, ill-treat, over-ride, over-drive, over-load, torture, infuriate or terrify any animal.

2. To cause unnecessary suffering by doing or omitting to do any act.
3. To convey an animal in a manner as to cause it unnecessary suffering.
4. To perform any operation without due care or humanity.
5. The fighting or baiting of any animal.
6. The administration of any poisonous or injurious drug to any animal.

It is worth noting that the Act specifically included mental as well as physical suffering, by including the words 'infuriate' and 'terrify'. The Act undoubtedly had a major effect on public attitudes, both by directly deterring abuse and also by changing people's perception of what was acceptable treatment of animals. The Act has been criticized (Todd, 1989) for using subjective terms such as 'unnecessary suffering' that make it difficult for a Court of Law to decide what is, and what is not, cruel. However, judges and juries must always interpret the law and, difficult or not, in hearing cases under this particular law they have found some defendants guilty and some not guilty. They have done so in the context of public attitudes: for a law to operate, it must be, to a greater or lesser extent, acceptable to the people it applies to, credible to the inspectors policing it and reasonable to those dealing with cases in the courts.

An important factor in the development of legislation on poultry welfare has been the variation in concern about animal welfare between European countries, mentioned in Section 7.3. Concern has historically been stronger in the north of Europe – particularly the UK, The Netherlands, Germany and Scandinavia – and weaker in the south. Reasons are complex. Several factors correlate with this variation, including temperature (hotter in the south, which affects how animals are kept) and religion (Catholicism is more common in the south, Protestantism in the north, with many effects on attitudes). The most persuasive explanation, however, is that concern has largely developed in urban people whose involvement with animals differed from that in rural areas – who kept pets more often than farm animals. The UK and The Netherlands, for example, were more industrialized than many other countries, and pressure for animal protection mostly came from city dwellers rather than from those involved in farming. A revealing snapshot was provided in 1981 by a review of which countries had then ratified the Council of Europe's 1976 Convention on the Protection of Animals kept for Farming Purposes (see Section 13.4). Of the 21 countries then members (Table 13.1), the 11 that ratified first were mostly from the north and had an average of only 6% of the population involved in agriculture. Switzerland is relatively southern but also relatively industrialized, and grouped with the north. Countries that ratified later had an average of 21% of the population involved in agriculture. They were mostly southern, plus for example Ireland with 23%. As well as differences in attitudes, this dichotomy may have reflected the fact that where many people are engaged in agriculture, their governments are unwilling to impose restrictions that affect their livelihood. Indeed, the agricultural industry has always been particularly vociferous and effective at lobbying for its interests.

In recent years, concern for animal welfare has also grown in southern Europe, as indicated by public opinion polls. In this case, however, actions of politicians have tended to lag behind changing attitudes rather than to promote them: southern

Table 13.1. Ratification of the Council of Europe's 1976 Convention on the Protection of Animals kept for Farming Purposes by 1981, and the proportion of each country's population involved in agriculture (Ludvigsen *et al.*, 1982).

Ratified	Agricultural labour (%)	Not yet ratified	Agricultural labour (%)
Belgium/Luxembourg	4	Austria	9
Cyprus	–	Greece	30
Denmark	8	Iceland	9
France	9	Ireland	23
The Netherlands	5	Italy	12
Norway	8	Liechtenstein	–
Sweden	5	Malta	5
Switzerland	5	Portugal	26
UK	2	Spain	17
West Germany	4	Turkey	54
Average	**6**	**Average**	**21**

European governments continue to be less positive towards welfare than northern ones (Sansolini, 1999).

13.3 Developments in Individual European Countries

Animal welfare legislation in individual European countries also shows a dichotomy that reflects attitudes. Northern countries have detailed laws, with codified lists of actions that are prohibited, rather like the UK's 1911 Act. Southern countries tend simply to state that animals must not be ill-treated. Legislation is also enforced more strictly in some countries than in others. We shall consider developments in individual countries first and then, in the next section, those in wider groups of European countries.

The most far-reaching legislation is the Tierschutzgesetz passed by the Federal German Parliament in 1972. It states that its basic principle is 'to protect the well-being of the animal. Without reasonable cause no one shall cause pain, suffering or injury to an animal.' The second part of the Act, which deals with the keeping of animals, says that a person who is keeping or looking after an animal: shall give the animal adequate food and care suitable for its species and must provide accommodation which takes account of its natural behaviour; and shall not permanently so restrict the needs of an animal of that species for movement and exercise that the animal is exposed to avoidable pain, suffering or injury. Furthermore, in 2002, the right of animals to due care was incorporated into the German constitution.

Denmark also has comprehensive legislation; for example, the Protection of Animals Act 1950, which states that: animals must be properly treated and must not by neglect, overstrain or in any other way be subject to unnecessary suffering; and anyone keeping animals should see that they have sufficient and suitable food and drink, and that they are properly cared for in suitable accommodation. This Act was interpreted as prohibiting battery cages, so there were no cages in Denmark for many years. However, Danish companies then started building farms over the border in Germany, and bringing the eggs back to Denmark. In 1979, a new Act allowed cages, but with a minimum area of 600 cm^2 per bird.

In the UK, public concern for welfare, including that of poultry, was further increased in 1964 by publication of Ruth Harrison's book *Animal Machines*. The Government set up the Brambell Committee, which reported in 1965 (Her Majesty's Stationery Office, 1965), passed the Agriculture (Miscellaneous Provisions) Act in 1968 and thence established an independent Farm Animal Welfare Council (FAWC). The Brambell Report and FAWC have had international impact, including through their development of the concept of Five Freedoms (Table 7.2) and through FAWC's reports (1986, 1991 and 1997). The 1968 Act also mandated production of Codes of Recommendation for the Welfare of Livestock (Ministry of Agriculture, Fisheries and Food, 1969). Contravention of these is not a legal offence in itself, but can be used as evidence in prosecution for cruelty (Chapter 7; Tables 7.1 and 7.3).

Sweden passed a new Animal Welfare Act in 1988, at a time when it perhaps did not expect to join the EU. This required that from 1989 all new cages for laying hens should provide 600 cm^2 per bird. Furthermore:

- Animals should be able to perform natural behaviours and be protected against disease and unnecessary suffering.
- Hens for egg production should not be kept in cages from 1999.
- But alternatives must not mean
  - impaired animal health;
  - increased medication;
  - introduction of beak trimming; or
  - impaired working environment.

However, by 1997, Sweden was a member of the EU. It was also argued that the latter conditions were incompatible with a complete ban on cages and the ban was deferred, subject to a requirement that all cages should be furnished (Section 12.6). A ban on cages remains on the statutes but in abeyance, and furnished cages were introduced in Sweden from 1998 (Tauson, 2000; Tauson and Holm, 2001).

Switzerland, which is not a member of the EU, banned cages for laying hens from 1992. This was the result of a referendum held in 1978, in which people were informed of the economic consequences of the decision. Not being in the EU, Switzerland can restrict imports of cheaper eggs; imports are needed, however, because the country is not self-sufficient in eggs. The law is framed as a ban on any enclosure for fewer than 40 birds. Various designs based on the Dutch tiered wire floor systems are used (Matter and Oester, 1989). It seems that performance of these, and welfare of the birds, was relatively poor at first but improved with experience (Fröhlich and Oester, 2001).

In addition to laws drawn up with the specific aim of improving welfare, there is legislation framed for other purposes that may have effects on poultry husbandry systems and management, and thus, indirectly, influence welfare. Relevant priorities and legislation also vary between countries. For example, Norway has legislation that limits the number of birds kept on a farm. Although its purpose is to encourage rural employment, by placing a ceiling on flock size it tends to discourage very

intensive husbandry systems. The Netherlands has laws on environmental impact
and pollution, that may discourage large-scale intensive systems by limiting the
quantities of waste material which can be produced.

Increasingly, however, legislation in European countries is subject to the wider
groupings to which we now turn. This includes legislation on employment, on the
environment and on animal welfare itself.

13.4 Council of Europe and European Union

The Council of Europe was started in 1949 to increase cooperation, and represents
most of the countries of Europe: the number rose to 44 in 2002. One area in which
the Council has been active is animal welfare, indeed it has stated that 'the humane
treatment of animals is one of the hallmarks of Western civilisation'. In 1976, it
produced the Convention on the Protection of Animals kept for Farming Purposes,
which was concerned with the care, husbandry and housing of farm animals,
especially those in intensive systems (Table 13.2). Its recommendations are couched
in general terms, but the drafting committee commented that they tried to lay down
principles precise enough to avoid a completely free interpretation, but wide enough
to allow for different requirements. Because the Convention itself is very broad, the
Council of Europe has a Standing Committee with a responsibility for elaborating
more specific requirements. One of the first areas in which it became active was that
of poultry welfare.

The Convention was not legally binding on member countries until they
ratified it; they then accepted the responsibility to include its provisions in their
national legislation. All the countries in Table 13.1 except Turkey have ratified by
2003, and others have also done so as they subsequently joined the Council. So from

Table 13.2. Key points of the Council of Europe's 1976 Convention on the
Protection of Animals kept for Farming Purposes.

Article 3 states: Animals shall be housed and provided with food, water and care
which – having regard for their species and to their degree of development,
adaptation and domestication – is appropriate to their physiological and ethological
needs, in accordance with established experience and scientific knowledge.

Article 4 states: The freedom of movement appropriate to an animal, having regard
to its species and in accordance with established experience and scientific
knowledge, shall not be restricted in such a manner as to cause it unnecessary
suffering or injury. Where an animal is continuously tethered or confined it shall be
given the space appropriate to its physiological and ethological needs.

Article 5 deals with lighting, temperature, humidity, air circulation, ventilation and
other environmental conditions such as gas concentration and noise intensity.

Article 6 deals with the provision of food and water.

Article 7 deals with inspection, both of the condition and state of the animal and of
the technical equipment and systems.

the late 1970s on, this Convention has been an underlying influence on legal developments in Europe.

The European Community started as a subset of the Council of Europe. It later expanded to 15 countries and formed the EU, and for simplicity we shall just refer to the EU here. It became a party to the Convention in 1978 and decided that it should act on the welfare of laying hens. First it produced Trading Standards Regulations for labelling of eggs (Table 12.1). Then, after several years of negotiation, a Directive was adopted in 1986 laying down minimum standards for the protection of hens in battery cages (Commission of the European Communities, 1986). By January 1988, all newly built cages had to provide 450 cm^2 per hen and other requirements (Table 13.3), and these standards were to apply to all cages by January 1995. Directives have to be translated into national legislation; in the case of the UK, for example, this was done in 1987 (Her Majesty's Stationery Office, 1987). Coupled with the main regulations that achieved this was a schedule setting out further requirements; these laid down specifications for cage design and construction, for daily provision of adequate and nutritious food and water, for good environmental standards including temperature and air quality, for a suitable diurnal lighting pattern, for thorough daily inspection of the birds, for prompt remedial action in the case of health or behavioural problems, and for the adequate functioning of automated equipment together with satisfactory back-up systems in cases of failure. However, the UK also amended its Welfare Code to recommend only the legal minimum of 450 cm^2 (Ministry of Agriculture, Fisheries and Food, 1987). Denmark and Sweden, in contrast, continued to provide more than this.

Meanwhile, national governments, particularly in northern Europe, and the EU, among others, financed background scientific work on poultry welfare and housing. The EU's 'Farm animal welfare co-ordination programme' dated from 1979. Politicians judged that public opinion – again particularly in the north – required more than the 1986 Directive. This was demonstrated, for example, by continued growth of the number of people who would pay more for non-cage eggs than for those from cages (Chapter 14).

Negotiations for a new Directive started in 1992 and were still continuing in 1998 when Sweden started introducing furnished cages. A new Directive was passed in 1999, and key provisions are shown in Table 13.4. It will phase out barren battery cages by 2012, with an interim measure requiring 550 cm^2 per hen from 2003. All

Table 13.3. Key points of the EU 1986 Directive laying down minimum standards for the protection of laying hens kept in battery cages (Commission of the European Communities, 1986).

A minimum area of 450 cm^2 per bird and 10 cm of feeding trough per bird

A continuous length of drinking trough providing at least 10 cm per bird or, if nipple drinkers or drinking cups are used, at least two shall be within reach of each cage

Cage height of at least 40 cm over 65% of the cage area and nowhere less than 35 cm

Cage floors capable of supporting adequately each forward-facing claw and not sloping more than 8°, unless constructed of other than rectangular wire mesh

Table 13.4. Key points of the EU 1999 Directive laying down minimum standards for the protection of laying hens (Commission of the European Communities, 1999).

Un-enriched (conventional) cages

From 1 January 2003, no new conventional cages may be brought into service and existing cages will have to provide 550 cm^2 per bird and a claw shortener

From 1 January 2012, conventional cages are prohibited

Enriched cages

From 1 January 2002, enriched cages must provide:

750 cm^2 per bird, of which at least 600 cm^2 is at least 45 cm high

A minimum total cage area of 2000 cm^2

A nest

Litter such that pecking and scratching are possible

15 cm perch per hen

12 cm of food trough per hen

A claw shortener

Alternative systems

From 1 January 2002, new non-cage systems must have:

A maximum of 9 hens/m^2 of usable area

Litter occupying at least one-third of the floor

15 cm perch per hen

From 1 January 2007, all non-cage systems must comply with these conditions

Review

By 1 January 2005, 'the Commission shall submit to the Council a report, drawn up on the basis of an opinion from the Scientific Veterinary Committee, on the various systems of rearing laying hens, and in particular on those covered by this Directive, taking account both of pathological, zootechnical, physiological, and ethological aspects of the various systems and of their health and environmental impact.

'That report shall also be drawn up on the basis of a study of the socioeconomic implications of the various systems and their effects on the Community's economic partners.

'In addition, it shall be accompanied by appropriate proposals taking into account the conclusions of the report and the outcome of the World Trade Organization negotiations.'

new cages from 2003 and all cages from 2012 must be furnished or enriched, providing 750 cm^2 per hen, a nest box, a perch and a litter area for scratching and pecking. Requirements for non-cage alternatives also change; litter is not currently required in percheries (Table 12.1) but from 2007 it will be needed in all houses.

The situation will be reviewed before the end of 2004, taking into account 'the socioeconomic implications of the various systems' and 'the outcome of the World Trade Organization negotiations'. How these factors and associated political developments are likely to affect implementation of the Directive will be discussed in Chapter 14. It is almost certain that there will be major changes to the housing of many or most laying hens in Europe over the next 10 years.

It is not possible here to review comprehensively the relevant legislation of all countries. However, we may note that just as with attitudes (Chapter 7), legislation has sometimes been affected by emigration from Europe. For example, Australia and New Zealand have legislation with some similarities to that of northern Europe. Canada also published a Code of Practice for the Care and Handling of Poultry (Agriculture Canada, 1989), which, though closely based in its approach and format on that produced by the UK Ministry of Agriculture, Fisheries and Food, transcends it both in content and in detail. For many areas, it provides sufficient information to serve as a basic handbook of good poultry husbandry. The European influence has not, though, been universal. For a contrasting picture, we turn to the USA, which, although also largely peopled by immigration from Europe, has very different legal protections for poultry.

13.5 USA

In the USA, statutory legislation dates back to 1641 when the Massachusetts Bay Colony framed their first legal code. Clause 92 read: 'No man shall exercise any tyranny or cruelty towards any brute creatures which are usually kept for man's use'. Their measure was far ahead of its time, for not until 1828 did one of the States of the Union – New York State – pass an anti-cruelty law. Most states now have such laws, allowing prosecution of animal cruelty as a felony in some states and as a misdemeanour in others. However, most exempt farm animals, either directly or for procedures that can be described as 'normal industry practices'. This means that practices such as high stocking densities and food withdrawal for forced moulting, which are not limited in any specific legislation and which are frequently more severe than in countries such as those of the EU, cannot currently be challenged as cruel. There are, however, some state laws that prohibit inhumane transport of poultry. Six states specify that crates for holding poultry shall conform to various requirements, often very modest. For example, Pennsylvania requires that live poultry shall not be stocked at more than 15 pounds per cubic foot; this is equivalent to about 240 kg/m^3. In Wisconsin, it is unlawful to transport chickens in coops unless the coops are 13 in (33 cm) high inside.

State laws are important, or potentially important, because in many ways the USA, as a union of semi-autonomous states, is parallel to the EU rather than to other, single countries. Also, whereas the EU now protects poultry, the USA does not do so. There are only three federal laws that apply to animal welfare. One has never been applied to poultry: the Twenty-Eight Hour Law of 1873 that limits length of animal transport across state lines. The other two specifically exclude poultry: the Humane Slaughter Act of 1958, and the Animal Welfare Act of 1966 that excludes all animals raised for food or fibre. This lack of legislation is surprising, as one analysis of the laws that do exist shows that the public conscience generally agrees that animals have the right to protection from cruel treatment, protection from abandonment, protection from poisoning and the provision of food, water and shelter (Animal Welfare Institute, 1991). It can be explained by a number of factors, such as concentration of animal agriculture in few states (and often far from main centres of population), domination of government agriculture committees by members from such states, powerful lobbying by the agricultural industry and a general reluctance to pass legislation limiting free enterprise (Garner, 1998). The lobbying power and industrial might of agribusiness and of other related sectors such as the fast food industry are now becoming a matter of national debate (Schlosser, 2001; Nestle, 2002). However, it also has to be said that since the year 2000, the retail sector has achieved more in improving how poultry are kept than has any legislation to date (Mench, 2003), as will be discussed in the next chapter.

13.6 Trade Organizations

The title of this chapter does not just refer to the actions of professional politicians but to all developments in policy and public affairs. Reference was just made to the influence of agribusiness. This is primarily exercised through trade associations, which in agriculture are able to recruit a high proportion of producers as members. Examples will be given from the UK and the USA. In the UK, the two main players are the National Farmers Union (NFU) and the British Egg Information Service (BEIS); there are smaller associations for categories of poultry other than laying hens, but producers of those categories are few in number. Both the NFU and BEIS have tended to resist pressure for changes intended to improve welfare. To some extent, they have modified their stance in recent years to reflect increased public concern for this issue. Booklets from both (British Egg Information Service, 1990; National Farmers Union, 1990) emphasize the importance of safeguarding birds' welfare. They also provide concise facts and summaries of the history, scope and present structure of the egg industry, including descriptions of the range of different production systems and setting out the advantages and disadvantages of each. The NFU publication has a short section on poultry meat, dealing primarily with broilers and also mentioning ducks and geese. However, the NFU continues to resist further change: when the European Directive on welfare of laying hens (Commission of the European Communities, 1999) was passed, their President wrote to the UK Minister of Agriculture in protest (Cruickshank, 1999).

In the USA, the equivalent of the NFU is the Farm Bureau, but the poultry trade associations are more influential: the United Egg Producers (UEP), the National Chicken Council (which deals with meat producers) and the National Turkey Federation. They have also tended to criticize calls for greater consideration of welfare. However, in about 1999, UEP started the process of drawing up detailed guidelines for their members on husbandry and welfare. The other associations have since followed suit, partly because from 2000, retailers started putting pressure on them to require humane treatment of animals. As these guidelines (United Egg Producers, 2002) mostly arise from economic pressure, they will be considered in the next chapter.

International trade-related associations have also tended to be conservative in this area. Following the 1999 European Directive, the International Egg Commission, representing 33 countries including all of the major producing countries except China, resolved to raise funding of US$1 million for action to overturn the ban on conventional laying cages. The resolution was supported by countries worldwide, including the USA. One reason must have been solidarity in the face of what was perceived as a direct attack on their European members, and in addition 'a domino effect is feared by the US, Canada and Australia' (Farrant, 1999, p. 1).

There are also trade associations for other groups which contribute to discussion and negotiation. Poultry scientists, for example, form societies such as the Poultry Science Association (in North America) and the World's Poultry Science Association. The latter has organized European Symposia on Poultry Welfare every 4 years since 1981 (following a predecessor in Denmark in 1977). The sixth was in Switzerland in September 2001. There has also been one North American Poultry Welfare Symposium (Mench and Duncan, 1998).

13.7 Animal Protection Societies

It was pointed out in Chapter 7 that many members of the public write to politicians about animal treatment, and this highlights the extent to which involvement in this subject is not confined to professionals. Among farm animals, a considerable proportion of this attention has been paid to poultry: 'Ban the Battery Cage' has been one of the most common of all protest calls, particularly in northern Europe. In the 20th century, it was probably surpassed as a popular cause only by a very few others such as 'Votes for Women', 'Ban the Bomb' and 'Save the Whale'. This activity is encouraged and coordinated by animal protection societies. Indeed, the widespread concern for the welfare of animals is reflected in the large numbers of societies and groups that have been set up in most countries. The core staff of these is generally professional, but they need to retain the support of their amateur supporters for their actions.

The emphasis of this volume has been on poultry welfare rather than animal rights (see Section 7.4). Most adherents of animal rights promote the argument that animal farming should be stopped altogether, but in Europe those who promote only this argument are few in number – although it may be pointed out that the case for abolition of animal farming is often part of the doctrine of some groups with other concerns, such as an emphasis on vegetarianism. Generally, support for animal rights in Europe is relatively slight so, of the animal protection societies that deal with poultry, most concentrate on the aim of improving the welfare of farmed poultry.

In the UK, for example, the RSPCA is a mainstream animal protection society that is active on behalf of all animals including livestock, and among many other activities lobbies for improved housing and conditions for farm animals. Dating from the early 19th century, it is the largest and most influential of such societies. Other societies also play important roles, especially as they tend to have special interests and concentrate their efforts on more specific issues. They include Compassion in World Farming, which campaigns for a ban on the export of live animals and is also very active on housing conditions. The Farm and Food Society is concerned with intensive husbandry, and the Farm Animal Care Trust works for improvements in the welfare of farm animals. There are also regional societies, such as the St Andrew Animal Fund and the Scottish SPCA, both based in Edinburgh. These societies are regarded as mainstream in having relatively moderate aims and using legal methods. They are generally well organized and pursue their objectives through lobbying Members of Parliament, circulating literature to their own members and to the public, writing to the press and placing paid advertisements.

Aware of the importance of influencing opinion at an early age, they also produce information packs and videotapes directed towards school teachers and children. They may thus exert an influence out of all proportion to the number of their members. The Universities Federation on Animal Welfare places a particular emphasis on education and dissemination of information, including the production of textbooks describing best practice (Ewbank *et al.*, 1999). It also funds research.

On the fringe of the welfare movement, there are other organizations, from non-violent activists such as hunt saboteurs to shadowy extremists in loosely organized groups such as the Animal Liberation Front who, as well as sometimes releasing farm animals, engage in activities involving damage to property and even threats to human life.

There are also, of course, animal protection societies in other European countries. Furthermore, there are international groups. One organization playing an increasing role in the EU is Eurogroup for Animal Welfare, which lobbies the EU's politicians on behalf of 15 member societies, one from each of the EU's member states. This is an important function, given the burgeoning part played by the EU in unifying and advancing animal welfare legislation. In addition, there are some societies that are active on the wider, world scene, such as the World Society for Protection of Animals and the International Fund for Animal Welfare.

In the USA, support for animal rights is stronger than in Europe, but there is also more overlap between concepts of welfare and rights. For example, Peter Singer is often referred to as the 'Father of the Animal Rights Movement' even though he is a utilitarian (Section 7.4) and therefore does not believe in rights. Some people use 'animal rights' as if belief in such rights were just a stronger form of concern for animal welfare (Ryder, 2000). Yet there is also tension between welfarists and rightists. Ryder describes the formation of new animal rights groups in the 1980s, including People for the Ethical Treatment of Animals (PETA). He then quotes Henry Spira, who was a pragmatic activist who worked on one issue at a time, as commenting that 'The war cry has been "all or nothing" with the almost inevitable result being nothing' (Ryder, 2000, p. 202). PETA certainly advocates veganism, animal rights and complete abolition of animal farming. However, in recent years, it has also campaigned for improved conditions of animals that continue to be farmed, including poultry.

Those who oppose pressure for greater animal protection, notably the poultry industry, have also perpetuated this blurring of the distinction between support for animal welfare and for animal rights. It is not uncommon, particularly in their public utterances, for them to describe all animal protectionists as 'animal rightists' and by implication extremists. Yet there are many societies that seek dialogue with the industry to negotiate change, while also lobbying for change through other routes. This approach started in the 19th century, as in Europe. The American Society for the Prevention of Cruelty to Animals was formed in 1866. Today, the largest US animal protection society is the Humane Society of the United States, with around 8 million supporters. Other groups such as the American Humane Association are also active in poultry welfare and will be mentioned again in the next chapter.

13.8 Influences on Attitudes

A theme of this chapter has been increasing public concern for animal welfare, including poultry welfare, over about the last two centuries. This increase in concern has both prompted, and been prompted by, political changes including legislation. Why has it occurred? One factor already mentioned is industrialization, with a decline in the number of people earning their living from farming. Several other factors are associated. Affluence in developed countries has increased, allowing people to diversify their interests beyond the question of whether they can afford the next meal. Perhaps most importantly, part of the industrialization process has been intensification of agriculture, and it is this that has provoked much of the concern for farm animal welfare. Worries are also expressed about other effects of intensive agriculture, such as on the environment. These issues were compounded in the UK and Europe, in the last years of the 20th and first years of the 21st centuries, by a succession of farm disease crises, including a major scare over *Salmonella* in eggs. The future of agriculture is receiving unprecedented attention, both in the UK with reports such as that by the Policy Commission on the Future of Farming and Food (2002) and throughout Europe with review of the Common Agricultural Policy. Anxiety about biosecurity has also increased internationally. On the issue of farm animal welfare, which correlates with many of these other issues, it seems likely that public concern will continue to wax rather than wane.

While this increase in concern has been general and uneven, there have been some specific triggers that have contributed, such as publicity from animal protection activists. Most notable in this regard were two books: *Animal Machines* by Ruth Harrison (1964) and *Animal Liberation* by Peter Singer (1975). Other influences are more ambivalent. Coverage of this issue in the media, in particular, does have some effect on public opinion – but must also reflect at least in part what the journalists and editors concerned consider of interest to the public.

One area that needs more study is how the attitudes of people involved in the poultry industry are changing as legislation and other influences on them change. We said in Section 7.3 that most farmers are concerned for their animals. It is our impression that although the industry as a whole may resist pressure for greater consideration of welfare, individual farmers and farm workers accept requirements for such consideration quite readily when they are actually imposed. For example, in the UK, the Codes of Recommendation for the Welfare of Livestock (Department of Environment, Food and Rural Affairs, 2002a,b) must be displayed in poultry houses and be familiar to the workers. Tours of poultry farms generally show that these requirements are met, and met without resentment. In the USA, the great majority of producers who are members of UEP have agreed to abide by the UEP Husbandry Standards (2002), including providing hens with a significantly increased amount of space, even though there is no legal requirement for them to comply and though there may be an increase in their cost of production.

The attitudes of the public, members of the poultry industry and others involved are primarily expressed, of course, within the economic context of agricultural production and sales. It is therefore to economics that we turn in the last chapter of this book.

14   Economics

14.1 Summary

- The economics of welfare is concerned with the implications for the economy of the whole food system, evaluating resources in agriculture, the effects on the real costs of food and how society's preferences are met. Welfare-friendly products tend to cost more; on the supply side, the producer has to take decisions regarding housing and husbandry systems.
- Production costs differ considerably between systems. Simple housing systems have lower capital costs but higher running costs. Loose housing systems for egg layers have greater food intake and higher labour costs than cages, as well as higher mortality.
- Stocking density has received considerable economic analysis. The economic/welfare trade-off of higher densities is finely balanced, and optimum profitability depends on relative egg and feed prices.
- Meat birds are mostly kept on the floor. Rapid growth, with low feed requirements, was standard until recently, but there is a growing market for free-range broilers and turkeys.
- Welfare considerations play an increasing role in the whole egg (but not egg products) market. In Europe, willingness of some people to pay more when they believe animals are being kept under better conditions has had important consequences – it has helped to drive the pressure for legislative changes intended to better animals' lives.
- Free trade threatens animal protection: welfare-friendly production in Europe might not be viable if it had to compete with imports from elsewhere where standards were lower. Voluntary agreements and emphasis on local origin may combat such free market pressures.
- Retailers increasingly recognize that customers are expecting them to offer welfare-friendly products. In the USA, major restaurant chains and supermarkets are producing husbandry guidelines for their suppliers.

- There is an emerging consensus that animal welfare is a public good, that it is best achieved by regulation rather than market forces and that this regulation can be exercised through voluntary agencies as well as by legislation.

14.2 Economic Attitudes

While some poultry are kept as companions, as ornamental birds or for fighting, most are kept for production, and this is the context of most of the consideration of poultry behaviour and welfare in this volume. We have emphasized this context, for example at the end of Chapter 13, because how commercial poultry are housed and treated will always be affected by monetary considerations – although one of our main points is that other considerations are also important. This final chapter will explore those monetary considerations, but it will also develop broader themes, because economics is not just concerned with money. Indeed, Bennett (1997, p. 235) explains that:

> Economics is not concerned with money *per se*, although it often finds money a useful measuring rod. Economics is concerned with how we in society make decisions about using resources to achieve the things that we want. The central problem for economics (and for society) is that we all want a multitude of different things but the resources that we have to achieve those wants are limited. We cannot therefore all have everything we want.

McInerney (1998, p. 115) develops the idea further in an article called 'The Economics of Welfare':

> There are two possible interpretations of 'the economics of animal welfare'. The popular conception is probably that it relates to the commercial implications of welfare improvements and the financial costs and benefits experienced by those directly concerned. On these grounds, one would expect to find statements about issues raised by higher welfare products . . .
> This 'supply side' viewpoint puts the emphasis on animal welfare adjustments as a problem for farmers, and treats economics as relating with the monetary side of things of primary concern to the producer. That, however, is an erroneously narrow view of economics – a view which reduces its role to one of simply measurement and (even worse) confuses the economist with an accountant.

McInerney's second, broader view of economics is similar to Bennett's and is considered in Section 14.8.

We must note that on either the narrow or the broad view of economics, human behaviour is very complex, and people attempting to understand it generally make assumptions that simplify the picture. Some of those assumptions turn out later, or in a different context, to be unreasonable. For example, Amartya Sen (1987, p. 12) points out that 'mainline economic theory' assumes that people will behave rationally, and predominantly defines rationality as 'maximization of self-interest'. Sen regards this definition as absurd, but it pervades much thinking about both economics and accountancy, among not only economists and accountants but also laypeople. We shall attempt to identify such assumptions in the discussions that follow.

At the risk of offending Bennett and McInerney, the next sections will consider primarily 'supply side economics', before the second half of the chapter turns to the broader mechanisms by which society has made decisions up to this point, and may do so in future.

14.3 Costs of Egg Production

The supply side economics of table egg production by laying hens has received more attention than that for any other area of agriculture, perhaps because such a variety of systems is available (Chapter 12) and the choice between them is so controversial. The comparison of costs among systems produced by Elson in 1985 is still useful (Table 14.1). He took the cost of production as 100 for birds housed in conventional laying cages at 450 cm^2 per bird. This was the minimum required in the EU until 2002, and at the upper range of allowances in the USA (Section 12.5). The EU requires 550 cm^2 per bird from 2003 (Section 13.4), close to the figure of 560 cm^2 included in the table. Greater space allowances in cages, as well as production in different systems, increase costs. Elson estimated that allowing birds 750 cm^2 in cages increases production costs by about 15%, housing on deep litter by about 18% and on free range by about 50%. The latter figure increases to 70% if the value of the land is included. However, land can often be shared with other stock such as sheep or cattle. Also, land usually appreciates in value rather than depreciating like housing and equipment. Other systems are intermediate in cost. Some capital items, such as high fencing to exclude predators, are expensive, and certain running costs, especially high food intake in winter, can be considerable. An extract of these results is also shown in Fig. 14.1.

Other estimates have differed from those by Elson. For example, Roberts and Farrar (1993) found that costs per dozen eggs produced were only 26% higher for free-range producers than for battery producers of a similar size. This difference demonstrates that such comparisons can never be definitive and will always be affected both by circumstances at the time of the study (such as the cost of certain inputs such as food) and by the particular parameters studied. One possible explanation for the relatively high cost of cage production in the later study is that it probably included only relatively small producers, so as to be comparable with the free-range producers. On the other hand, comparing small free-range producers with large cage producers would have been questionable on other grounds. There is no 'correct' comparison possible here. At any rate, if systems are ranked according to their costs in Table 14.1, this ranking will be correct or approximately correct in any reasonable circumstances.

A more recent comparison of different systems and countries is shown in Table 14.2. Here even more assumptions had to be made, but again the broad result – that costs vary between systems, and even more so between countries – appears sufficiently robust not to depend very precisely on those assumptions.

Table 14.1. Egg production costs in different systems for laying hens, relative to laying cages with 450 cm² per bird. Space refers in cages to cage floor area, in houses to house floor area and in extensive systems to land area (Elson, 1985).

System	Space	Relative cost (%)
Laying cage	450 cm² per bird	100
Laying cage	560 cm² per bird	105
Laying cage	750 cm² per bird	115
Laying cage	450 cm² per bird + perch	100
Laying cage	450 cm² per bird + perch + nest	100
Shallow laying cage	450 cm² per bird	102
Get-away cage, two-tier aviary	10–12 birds/m²	115
Aviary	10–12 birds/m²	115
Aviary and perchery and multi-tier housing	20 birds/m²	105–108
Deep litter	7–10 birds/m²	118
Strawyard	3 birds/m²	130
Semi-intensive	1000 birds/ha	135 (140[a])
Free range	400 birds/ha	150 (170 [a])

[a]Includes land rental.

In examining costs of production, it is necessary to consider the factors that influence them. These include housing, labour, food intake, hygiene, mortality and predictability of performance.

The simpler housing systems tend to be lower in construction costs. For example, the strawyard system is uninsulated and has no fans and may be a cheap pole barn construction. However, stocking density is very low in such a system (3–5 birds/m²) and therefore housing cost per bird is higher than in more densely stocked systems.

Labour is an important cost factor. Loose housing systems such as percheries and aviaries require more husbandry skill and a greater labour input. For example, there are floor eggs to collect, birds to be trained to use nest boxes, treatment of pecked birds, nest boxes to be closed at night, and so on. Dedicated staff, often prepared to work long hours in less favourable conditions, are required, which means increased labour costs. In a comparison between three systems by Tucker (1989), labour costs in UK£ per dozen eggs were 0.015 for cages, 0.032 for a perchery and 0.133 for free range.

Food intake is the largest component of egg production costs and varies widely both between and within systems. The three main factors in this variation are temperature, activity and feeding space. Within housing, food intake generally increases with lower stocking density, because the latter tends to result in lower house temperature, which increases the food requirement for maintenance (Section 10.6). So food intake is higher in non-cage systems than in cages. A similar but smaller effect of density also occurs in cages, e.g. Roush *et al.* (1984) found that hens stocked at three per cage ate 30.1 kg of food each in the laying year, while those at five per cage ate 29.4 kg. This was probably mediated by all three factors of temperature, activity (with birds at lower stocking density more active) and feeding

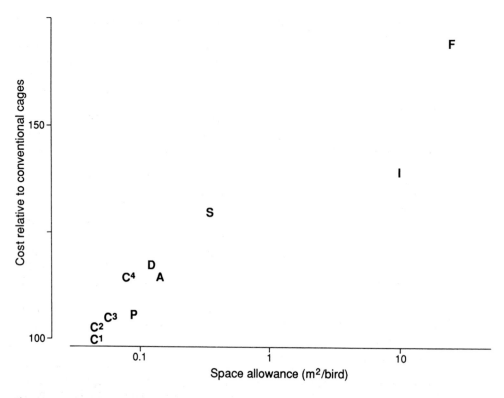

Fig. 14.1. Cost of egg production in different systems (simplified from Table 5.1). Space allowance is plotted on a logarithmic scale. C^1, 450 cm^2 cages; C^2, shallow cages; C^3, 560 cm^2 cages; C^4, 750 cm^2 cages; P, perchery; D, deep litter; A, aviary; S, strawyard; I, semi-intensive; F, free range (Appleby and Hughes, 1991).

space. Temperature was ruled out in another study which also found that birds ate more when stocked fewer per cage, compared with those at higher stocking density within the same house (Appleby *et al.*, 2002). However, both temperature and activity are implicated in higher food consumption by hens allowed outdoor access. In Tucker's study, food intake in g/bird/day was 115 in cages, 116 in the perchery and 135 in free range. In addition, fewer eggs were collected in the latter systems, so food costs in UK£ per dozen eggs were 0.26, 0.28 and 0.33, respectively. Together with labour costs, these figures largely accounted for differences in the egg production costs of the three systems.

It is not clear whether there are any welfare implications of higher or lower food intake, as even high stocking density probably does not prevent birds from getting to feeders at some time. However, high density does cause other problems, so it is important to note that its commercial advantage is not as inarguable as generally assumed. Increasing the size of a laying flock by placing more birds in each cage produces more eggs (but fewer per hen and of lower weight) but also, of course, requires more food. So the profitability of this depends on food costs and egg prices, and to a lesser extent on other costs such as that of purchasing pullets. Roush *et al.* (1984) showed that profit was greater with fewer birds per cage than the

Table 14.2. Estimates of egg production costs in different systems and countries: US$ per dozen eggs (at $1 = €1.07 = Sfr1.68).

	USA	EC			Switzerland	
	Caged	Caged	Barn	Free range	Barn	Free range
Variable costs						
Food	0.24	0.29	0.33	0.40	0.79	0.81
Health care		0.01	0.01	0.01	0.02	0.02
Miscellaneous	0.03	0.02	0.03	0.02	0.03	0.03
Bird depreciation	0.05	0.11	0.13	0.14	0.45	0.45
Total	0.33	0.44	0.49	0.58	1.28	1.30
Overheads						
Labour	0.04	0.04	0.08	0.20	0.28	0.29
Electricity		0.01	0.01	0.02	0.03	0.03
Water		0.00	0.00	0.00	0.01	0.01
Total	0.04	0.06	0.10	0.23	0.31	0.33
Total cash cost	0.37	0.49	0.59	0.80	1.60	1.63
Housing and equipment	0.05	0.09	0.13	0.12	0.35	0.41
Total cost	**0.42**	**0.58**	**0.72**	**0.92**	**1.95**	**2.04**
Total cost 2003		0.63				
Total cost 2012		0.73				

Assumptions include space allowances (350 cm^2 per hen in US cages, 450 cm^2 in EC cages, 25 hens/m^2 in EC barns, 7 hens/m^2 in Swiss barns) and other variables such as egg production and cost of feed. Costs for free range do not include land rental. Estimates of total costs are also given for EC cages from 2003, providing 550 cm^2 per hen, and from 2012, when they will be enriched (Table 13.4). (After Fisher and Bowles, 2002.)

maximum over quite a wide range of economic conditions. Yet Bell and Swanson (1975) have indicated that it is difficult for some poultry managers to visualize that fewer hens can make more revenue than a more crowded flock.

Hygiene is important for food safety and, since in many non-cage systems the birds are not separated from their droppings, care needs to be taken in them to avoid soiling of eggs – which reduces income – and of other birds. Contact with droppings also increases risk of certain infections and parasitic infestations (e.g. coccidia and ascaridia). These cause economic losses through reduced production and because eggs cannot be marketed during and following treatment if drugs have to be used that may leave residues in eggs. In addition, there may be increased production costs due to extra staff vigilance to observe warning signs, labour to keep litter in good condition, medication and veterinary costs when necessary. There is also greater risk of certain external parasites, e.g. red mite, in some cage-free systems. These also increase production costs through lowering birds' performance and increasing need for treatment.

The death rate in birds housed in non-cage systems is generally, although not necessarily, higher than in caged layers, and includes deaths from cannibalism. It can also include predation on range. Several other factors in such systems combine to make performance less predictable, thus increasing the risk of high production costs. One of these is the risk of floor eggs, which are more costly to collect and are often dirty or cracked and therefore downgraded. Another is the effect of weather conditions on the stock (temperature, humidity and wind), which can be much greater and less controllable particularly where stocking densities are low.

Risk is an important factor in decision making by poultry managers. Roush (1986) incorporates various possible approaches to risk in his analysis of decisions about how many laying hens to place in each cage, given that food purchase costs and egg selling prices may change during the year. For example, under a Maximin approach, the manager would estimate the minimum likely return for each stocking density, and select the highest of these. A Maximax approach, however, focuses on optimism, and selects the density that may give the highest possible return. Roush comments (1986, p. 30) that:

> It is accepted in decision theory that two responsible managers might choose different optimal actions. For example, the poultry manager who is attempting to become established and who does not have the same ability to withstand an unfavourable outcome as a more established poultry manager may consider the conservative Maximin approach to decision making.

Roush suggests that managers should adopt a decision analysis approach for their own profit, and it is interesting that this supports his earlier point (Roush *et al.*, 1984) that managers would sometimes benefit from choosing lower stocking densities. This would also give a benefit, of course, for the welfare of the birds concerned.

14.4 Costs of Meat Production

Most broilers, turkeys and ducks are loose housed on litter and reared as rapidly as possible to obtain maximum growth rate and food conversion. In some Eastern European countries and in the former Soviet Union, many broilers are grown in cages. Capital costs are higher for cages than for floor systems, but these may be partly offset by savings in running costs, with no need for litter and less need for heating houses because of higher numbers of birds in a house of the same volume.

On a smaller scale, an alternative approach is developing, especially in France and the UK under the trade name Label Rouge but also to some extent in the USA, for slower growing, free-range broilers and turkeys. Current indications suggest that free-range poultry meat production will expand rather like free-range egg production has in Europe. Housing and production costs are considerably higher, because of lower stocking density within the building, the fact that after 3 weeks of age the birds have access to the outside and the fact that birds are kept for longer. Food consumption is higher, both on a daily basis and because it takes longer for birds to reach selling weight, and food conversion is therefore less economic. However, selling prices are higher, either through supermarkets or through small-scale outlets.

Two other approaches that overlap with the speciality production just mentioned are organic production and welfare-labelled products. Organic standards are maintained in the UK by the Soil Association, and in the USA (from October 2002) by the National Organic Standards Board. Probably their most important elements are that birds to be sold as organic should be raised on organic feed and without synthetic drugs. However, the organizations drawing up and supervising the standards have recognized that both consumers and most organic producers expect livestock to have outdoor access and to have reasonable standards of welfare. As such, costs of organic production are also higher than normal commercial costs. Similarly, meat products sold as having met specified welfare standards may have to come from birds that have been outdoors.

14.5 Egg Market

Even 'supply side economics' is not limited to the supply side, of course, but involves interaction between supply and demand, between producers, retailers and customers. This interaction will be discussed here with particular reference to hens' eggs. Welfare considerations have played more part in egg sales than in any other sector of poultry production. Indeed, sales of eggs from systems such as free range have led the way for welfare improvements in all livestock production. We must note at the outset, however, that this applies mostly to eggs sold whole. Few ready-made meals or other products containing eggs indicate how the hens were kept (an exception may be organic products), and few customers think to ask – although commercial purchasers may do so, as discussed below. The fact that an increasing proportion of food is sold in processed form will also be considered below.

Welfare considerations aside, the egg market tends to have cycles of overproduction and low prices alternating with relative shortages and higher prices. The latter tend to be short lived, because production can be stepped up fairly rapidly and because when prices are high, eggs are imported from regions or countries that have a surplus, if that is permitted. The egg market and selling prices are organized slightly differently in various countries. Prices are to a large extent controlled by supply and demand, but there are market leaders who tend to set the price, which is then held if other traders follow. In The Netherlands, there is a weekly auction that sets the price. In the UK, market leaders meet once a week to set a guide price. Then prices tend to be fairly uniform within countries over short periods. Most countries now have a free egg market and, in these, despite the efforts of the poultry industry to maintain them at what it regards as acceptable levels, egg prices are generally low in comparison with most other protein foods. Certain countries operate quota systems that control egg output and, in these, egg prices are generally higher, giving producers more reliable margins.

The demand for eggs is inelastic: eggs are not readily interchangeable with other items in the diet and people tend to buy a set number whatever the price. Lower prices do not generally result in increased sales and, conversely, higher prices do not reduce sales. Indeed, there have always been some consumers who prefer to pay more for eggs, apparently with the feeling that by doing so they are getting a better-quality product. This preference had a strong influence in the UK in the 1960s. Until then, most eggs were white, from light hybrid hens, as they still are in most other parts of the world. However, people bought brown eggs when they were available, even though they cost more, because they were perceived as tastier or more natural than white ones. By the 1970s, most commercial production in the

UK had changed over to using medium hybrids, producing brown eggs. People are also sensitive to the quality of eggs. Shells must be clean and intact, yolk colour deep enough to satisfy demand in the area, and albumen firm enough to ensure that eggs stand up well when broken out.

This background may explain why, uniquely among animal production sectors, the system in which eggs were produced became a selling point. A niche market began to develop, particularly in northern Europe, for eggs that did not come from cages: free-range eggs in some countries, deep litter or 'scratching' eggs in others. Initially this was mostly roadside sales in the country, from farmyard flocks, but in the 1980s such eggs also began to be produced by larger companies and sold in shops. Some people bought them – at a higher price – because they perceived them to be more nutritious, tastier or healthier. Some were also concerned about the welfare of the hens, and this concern led to the development of other non-cage systems in the 1970s and 1980s (Chapter 12). A problem was that eggs sold as free range might come from hens allowed to range only inside a house, or only if they could find one small exit from a large building, and other terms were similarly ill defined. The EU acted in 1985 by imposing the Regulation discussed in Chapter 12, defining four labels that can be put on eggs and the corresponding conditions in which hens must be kept (Table 12.1). By the 1990s, about 20% of eggs sold in the UK were free range, and an additional proportion from percheries (sold as barn eggs). In the late 1990s, some supermarket chains actively promoted this trend. Some sold barn eggs at the same price as cage eggs, while one chain stopped stocking cage eggs altogether. In other countries such as Denmark and Germany, deep litter eggs are more popular.

A similar trend began in North America in the 1990s, but is still small to date. There terminology varies again, with eggs being sold with descriptions such as 'cage free'.

When free-range eggs were a rare speciality, sold by places such as health food shops, they could be sold at up to twice the price of cage eggs, or even more than this. This pricing differential then tended to be retained even in supermarkets, despite the fact that production costs were only 25–70% more than cage eggs (see above). In other words, packers or retailers claimed a higher profit on these eggs. This practice has been criticized (Harrison, 1991) on the grounds that it probably restricts sales and prevents a larger shift towards free-range production. However, it persists to the present day, although with considerable variation. Litter and barn eggs are also generally priced higher than cage eggs: anything from 15 to 50% higher.

It remains questionable whether many consumers know what non-cage systems are like: for example, they are probably unaware that most free-range hens are not in farmyards but in very large flocks, and that 'barns' do not have to have litter on the floor (Table 12.1). Nevertheless, the fact that so many people were prepared to pay more for eggs that they believed were associated with higher welfare was important. It was probably one of the most important factors that led European governments and the EU to legislate for improved hen welfare, leading up to the 1999 Directive phasing out conventional laying cages (Sections 12.5 and 12.6). Indeed, it led the way for improved welfare of all farm animals as social, economic and legislative pressure for changes in the treatment of farm mammals tended to follow those for the treatment of poultry.

14.6 Competition and Trade

The last two sections concentrated on minority sales of meat and eggs from alternative systems. The fact remains that the majority of poultry is kept in intensive systems. Intensification was initiated by public policies – before, during and after the Second World War – in favour of more abundant, cheaper food. As a result, animal production became much more efficient, as measured by the cost of producing each egg or kg of meat. The pressure for efficiency subsequently became market driven, with competition between producers and between retailers to sell food as cheaply as possible, and thereby acquired its own momentum. In practical terms, the effects have been spectacular. In the post-war period, a meat bird took over 13 weeks to grow to 2 kg and cost about UK£40 in today's money. Today, it takes less than 6 weeks (Fig. 14.2) and costs less than UK£2. Eggs cost about one-eighth of their price 50 years ago, in real terms.

As part of this competition, farms have increased in size and keep more birds per unit area, either in housing or on pasture. Methods used on farms have favoured automation and other approaches to reduce labour needed for each animal, e.g. use of cages that control the behaviour of the birds to make their management easier. Birds have also been bred to produce meat or eggs faster and with a relatively lower input of feed. So these changes in agriculture have had a price, and to a great extent that price has been paid by the animals. They typically get less space per individual than they did previously and many live in barren environments that do not allow

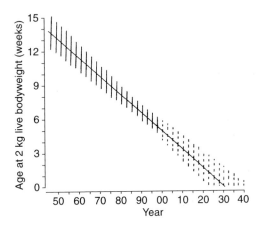

Fig. 14.2. From 1950 to 2000, the time needed for a broiler chicken to grow to 2 kg was reduced steadily, as shown by the sloping line (the vertical lines show maximum and minimum values). If similar reduction was possible in future, as shown by the part of the graph with dashed vertical lines, chicks would eventually be 2 kg just after hatching. However, there must be limitations that will soon prevent further reduction (Etches, 1996).

them to exercise their normal range of behaviour, while genetic selection has been accompanied by increased problems with production-related diseases. Profits from increased efficiency are generally short term, as they are regularly pared away by competition to reduce selling prices. Yet some of the changes by which efficiency has been increased, for example reduction in space allowances, have produced long-term reductions in welfare. However, in 2003 the World Organization for Animal Health (OIE) began procedures to establish international standards for animal welfare.

There are also what may be called major side effects of low selling prices for food. The cheapness of broilers has made disposal of end-of-lay hens, previously sold for stewing or soups, a serious problem (Section 11.14). In the USA, it has reached such a point that there are difficulties in selling dark meat from broilers, as customers prefer white meat and reducing the price of dark meat further is insufficient incentive for people to buy it. This was doubtless part of the motivation for a development in which very large numbers of chicken legs were donated to Russia as food aid in the early 1990s; once consumption patterns were established, subsequent consignments were sold. By 2002, the broiler industry had sufficient income from the trade that when Russia temporarily refused entry to this product, this was a matter of significant financial concern.

Indeed, low selling prices are a regular cause for concern throughout the industry, despite the major role that the industry has played in reducing those prices. Ironically, the industry both boasts about the cheapness of the food it sells, and complains of the low prices it receives. This is especially a problem for smaller companies – and indeed the main reason why so many have gone out of business – but insecure income is also an issue for the big companies. Understandably, they wish to assure future income. This is why the argument has strengthened in recent years, in the main poultry-producing countries, that export of poultry products is vital to the industry. When something interrupts such exports, such as Russia's embargo on their importation, it is commonly said that this is a severe threat to an industry already struggling to survive. This would once have been incredible: it should obviously be possible for a large food-producing industry to be self-sufficient in an affluent country such as the USA. It remains only partly true: a significant counter-argument is that in many such countries, imports of poultry products approximately balance exports (Lucas, 2001). However, it is now credible, simply because the increase in efficiency and the reduced selling prices have made margins so tight that finances of poultry companies are constantly on a knife edge. Some companies certainly make large profits. However, even large companies can rapidly get into difficulties; one such, owning over one million hens in the southeastern USA, went bankrupt in 2002.

The situation is getting worse, partly because trade itself increases competition and pushes prices down further. This also exacerbates the difficulty of addressing issues such as animal welfare and environmental protection within countries. Yet increased international trade in agricultural products is clearly regarded as desirable by governments and the agricultural industry. As such, negotiations are under way to extend the rules for free trade established by the World Trade Organization (WTO) to agricultural products.

Discussions in the WTO are already raising questions about implementation of the 1999 EU Directive phasing out conventional laying cages (mentioned above and covered in Chapter 12). Indeed, the Directive includes the requirement for a review, by January 2005, that will consider performance of different systems and their 'socioeconomic implications', together with 'the outcome of the World Trade Organization negotiations'. The EU proposes that animal welfare should be taken into account in trade, by allowing either agreements between trading partners that safeguard welfare, or labelling, or payment of subsidies to producers who maintain high welfare standards (European Communities, 2000). This is meeting resistance from other countries including the USA.

The EU's Scientific Veterinary Committee (1996, p. 111) suggested that within the EU, 'High standards of laying hen welfare can only be implemented and sustained if the EU market is protected against imports of eggs from third countries with lower standards'. That view is supported by a suggestion that, in its current form, the Directive will weaken EU competitiveness to such a degree that 65% of domestic consumption could be substituted by imported eggs (Wolffram *et al.*, 2002), given the fact that other countries can produce eggs more cheaply (Table 14.2). However, it can be argued that this is a considerable overstatement. Denmark has had more stringent legislation on cages than the rest of Europe for years. Its egg industry survives, albeit perhaps smaller than it might otherwise have been. If this applies within Europe, it applies even more to the threat of longer-distance imports to European countries from outside Europe, at least with regard to whole eggs. There is a danger, however, that imports of processed eggs, which make up 25% of European egg production, would rise in the absence of protection.

14.7 Stakeholders

The interactions between producers, retailers and consumers in the egg market raise the broader question of who the stakeholders are in decisions about poultry welfare, and of what contribution they have made to those decisions. To those three categories can be added others considered in Chapter 13, e.g. legislators, media and people active on welfare issues. Categories of people overlap. In particular, everyone is a consumer, and it is also worth emphasizing that everyone is a citizen. However, those with special interests will necessarily have different priorities from others. At least three other categories of stakeholder could also be listed. First, there are groups of people, such as communities, whose interests may be considered in distinction from those of individual people. Secondly, there is the environment, although in so far as the environment can be said to have interests these will most readily be represented by those people with particular concern for such issues. Finally, there are the birds themselves.

The previous section discussed the pressure for cheap food production that has been widespread over the last 50 years or so. This pressure is sometimes described – including by the poultry industry – as a consumer demand for cheap food, but this is an oversimplification implying that people want cheapness at the expense of all other considerations and that cutting prices is an end that justifies all possible means. It is not surprising, indeed it is reasonable, that offered two otherwise similar products most shoppers will buy the cheaper. Surveys have shown that more people say they want welfare of farm animals to be improved, even if this increases food prices, than actually buy higher-priced welfare-friendly products such as free-range eggs in the shops (Bennett, 1997). This is still reasonable, because they are behaving as citizens when they answer the questionnaire, and as consumers juggling varied priorities when they do the shopping. In the only case where people have actually been asked to vote on legislation to improve animal welfare, with associated higher costs, they approved that legislation: this case was in Switzerland, where battery cages were banned by referendum (Section 13.3).

In fact it is not reasonable to expect shoppers to take day-to-day responsibility for animal welfare at the point of sale – any more than they are expected to do so for other issues that are of concern to society such as pollution. It is increasingly apparent that people who do not look after farm animals themselves expect those who do to take responsibility for doing so properly – either voluntarily or involuntarily. This expectation is being realized in both Europe and the USA, but by different mechanisms (Mench, 2003). In Europe, the social attitudes that were indicated by the increasing number of people buying free-range eggs and other welfare-friendly products are now being translated into legislation that increases safeguards for animal welfare. Furthermore, while the initiative started with poultry welfare, it then extended to welfare of all farm animals. In the USA, the lead has largely been taken by the retail sector. A senior executive of one of the major fast food chains has commented that their customers expect them – the restaurant company – to ensure that the animals supplying them with food are properly looked after (England, 2002). As such, in 2000, another of those companies, the McDonalds Corporation, which buys 2.5% of US eggs, started requiring its suppliers to provide laying hens with the same space allowance as in Europe, and not to practise forced moulting by withdrawing feed. Other companies followed them in drawing up their own requirements, and the National Council of Chain Restaurants and the Food Marketing Institute (which represents the major supermarket chains) began approving husbandry guidelines for their suppliers of animal products in 2002.

In the case of laying hens, those guidelines have basically endorsed the husbandry standards of the trade association, the United Egg Producers (Section 13.8). The National Chicken Council and the National Turkey Federation have also outlined standards. These do not go as far as European legislation, but they are important in acknowledging the importance of poultry welfare, and in forming a basis for possible future raising of welfare standards.

The part played in all this by people particularly concerned about poultry welfare includes the buying of products perceived as beneficial for welfare, such as free-range eggs and meat. It also includes the activities of the animal protection societies, both in putting pressure on other groups and, more specifically, in launching schemes that directly address concern for animal welfare. The leader in this field is the UK's RSPCA, which launched its Freedom Food programme in 1994. There are detailed criteria that must be met by producers who want to join the programme. They can then use the Freedom Food label. This includes the name 'RSPCA', which has widespread recognition and confidence from the British public. The RSPCA also helps with marketing. The programme has grown steadily, helped by the overlap in criteria between this and other schemes. If producers are already certified organic or for producing free-range eggs, they do not usually have to make many additional changes to be able to use the Freedom Food label, which is therefore well worthwhile. Similar programmes have begun in North America. The American Humane Association started its Free Farmed scheme in 2000, Certified Humane was launched by Humane Farm Animal Care in 2003, and in Canada SPCA-certified food was launched in British Columbia in 2002.

As we have said, one final category of stakeholders in decisions that affect poultry welfare is the poultry themselves. Poultry do not speak for themselves in this, except in so far as the evidence that can be obtained about their preferences and needs, presented in Part C, can be taken as pertinent communication. However, most people who give attention to this subject accept that our relationship with the animals we keep is two-sided, not just one-sided *use* of animals without restriction. Since we are in control of that relationship, it is our responsibility to look after both sides, not just our own. The two-sided nature of the relationship is sometimes expressed in terms of an implicit contract between humans and animals. They help us, so we look after them.

Poultry production is at base a biological process, not a technological one, and it is important to remember that poultry are birds, not machines or humans or mere economic units (Section 7.6). How to retain that understanding in practice is a question for the future of agriculture and of society.

14.8 Making Decisions

The previous section began to consider the broad mechanisms by which society has made decisions up to this point. This is McInerney's second and preferred understanding of welfare economics (1998, p. 116):

> The more relevant interpretation is that 'the economics of welfare' examines the implications for the economy of the whole food system, including the way resources are used in agriculture, the effects on the real costs of food and the extent to which the outcome meets the preferences of society.

The fact that the retail sector takes into account the opinions of its customers is part of this broad picture, but it remains to be seen how far that process goes in the USA. McInerney's view (p. 123) is that:

> The mechanisms for articulating specific preferences back down the complex geography of the food chain are in general too crude (or non-existent) to make overall animal welfare standards except in the most limited instances (battery eggs, organic meat) susceptible to market choice.

This is partly because what he calls 'the real costs of food' are very involved: they include many costs that are not paid by producers but by society (externalities), and many that are difficult to assess in monetary terms, such as the effects of the production system on the environment, food safety and quality, food security, family farms, farm workers, rural communities and developing countries. It is unlikely that decisions by retail companies can ever fully address such concerns, and McInerney concludes (p. 124) that 'If animal welfare is a public good, regulation (not market forces) has to determine standards'. This is what has happened with the legislative process in Europe, although a problem still to be faced is that in the context of increasing globalization such legislation has to take public opinion into account not just in Europe but worldwide (Section 14.6). Regulation does not necessarily have to be by legislation, as self-regulation by industries is also possible, but self-regulation is variable in its effectiveness and in its responsiveness to public opinion.

Control by regulation, taking public opinion into account, avoids the limitations inherent in 'purchasing power', e.g. the tendency of the latter to apply to whole eggs but not to egg products. This is important because an increasing proportion of food is sold in processed form.

In fact the shift towards sale of pre-processed food in developed countries offers hope for widespread improvement of farm animal welfare. If a meal containing animal products is bought in a supermarket or restaurant, those products account for only about 5% of the price. So an increase in cost of animal production by, say, 10% would only increase the cost of such meals by 0.5%. Most customers would not notice such a change and would approve it if asked, to benefit animal welfare or the environment.

McInerney (1998) goes on to analyse the financial impact of banning certain livestock systems. He estimates that banning intensive rearing of broilers, and keeping them in smaller groups, with larger areas and more varied conditions, would increase meat production costs by 30%. However, such a ban would increase retail prices by only 13%, because these prices include transport, packing, marketing, and so on: changes in production costs are 'diluted' by the further costs of bringing products to market, plus the 'mark-up' added by retailers. Similarly, banning battery cages would increase farm costs by 28% but retail costs by only 18%. Neither broiler meat nor eggs are a major component of household food expenditure, so the effect of both measures together would be to add about UK£0.06 to weekly food bills. Meanwhile, it should be possible for the farmers to maintain their profits, offsetting increased costs with increased selling prices. These calculations illustrate two general propositions (McInerney, 1998, pp. 127, 130):

> The economic costs of reasonable improvements in animal welfare are likely to be relatively small;

> Higher animal welfare standards are not an economic imposition on farmers.

An obstacle to such change, however, is what may be called economic inertia. Producers tend to resist legislation – or pressure from intermediary buyers – to improve conditions for animals because in existing price structures buyers continue to expect low prices. Any increased cost of production would therefore tend to be borne by producers and they would suffer losses or reduced profits, at least in the short term. If these short-term effects can be avoided, though, by making changes gradually or providing public subsidy, a new situation with increased costs and increased income from increased food prices need not be disadvantageous to producers.

Major questions remain about how decisions affecting animal welfare can be made better in future, taking into account the needs of all the stakeholders involved. Leaving such decisions to the 'free market' is no longer adequate, especially given the disproportionate power in such a market accrued by the biggest players such as large agribusiness and retail companies. For example, it can be argued that competition should no longer be the main determinant of food prices, where these affect major issues of concern to society, notably the environment and animal welfare.

One thing is clear: that decisions about the structure of agriculture will in future have to take greater account of public opinion than hitherto. This should not be a burden on farmers, who are stewards on behalf of society of the animals that feed us and of our environment. On the contrary, it should ensure farmers a more valued place in society and a more reliable income. The importance of public opinion is emphasized in numerous discussions of agriculture – especially in view of what is widely regarded as a crisis in European agriculture in the early 21st century (Section 13.8). Thus one of the main recommendations of the UK's Policy Commission on the Future of Farming and Food (2002) is greater integration and communication between all stages of the 'food chain', from producer to consumer. Also, in the USA, the National Research Council (2002), reviewing the research programme of the US Department of Agriculture, recommends increased public accountability, for example by holding a public discussion forum every 2 years. It also – to return to our main point of changing priorities – recommends that government-funded research should in future be less on productivity and more on public goods such as environmental stewardship.

We need a vision for agriculture in the future that will be sustainable for the animals that feed us, for our environment and for ourselves. Safeguarding poultry welfare should be a part of that vision.

References

Abrahamsson, P. and Tauson, R. (1993) Effects of perches at different positions in conventional cages for laying hens of two different strains. *Acta Agriculturae Scandinavica* 43, 228–235.

Abrahamsson, P. and Tauson, R. (1997) Effects of group size on performance, health, and birds' use of facilities in furnished cages for laying hens. *Acta Agriculturae Scandinavica* 47, 254–260.

Abrahamsson, P., Tauson, R. and Appleby, M.C. (1995) Performance of four hybrids of laying hens in modified and conventional cages. *Acta Agriculturae Scandinavica* 45, 286–296.

Abrahamsson, P., Tauson, R. and Appleby, M.C. (1996) Behaviour, health and integument of four hybrids of laying hens in modified and conventional cages. *British Poultry Science* 37, 521–540.

Adams, A.W. and Craig, J.V. (1985) Effects of crowding and cage shape on productivity and profitability of caged layers: a survey. *Poultry Science* 64, 238–242.

Aerni, V., El-Lethey, H. and Wechsler, B. (2000) Effect of foraging material and food form on feather pecking in laying hens. *British Poultry Science* 41, 16–21.

Agriculture Canada (1989) *Recommended Code of Practice for the Care and Handling of Poultry from Hatchery to Processing Plant.* Publication 1757E. Agriculture Canada, Ottawa, Canada.

Aho, P.W. (2002a) The world's commercial meat and chicken industries. In: Bell, D.D. and Weaver, W.D. (eds) *Commercial Chicken Meat and Egg Production,* 5th edn. Kluwer, Dordrecht, The Netherlands, pp. 3–17.

Aho, P.W. (2002b) Introduction to the U.S. chicken meat industry. In: Bell, D.D. and Weaver, W.D. (eds) *Commercial Chicken Meat and Egg Production,* 5th edn. Kluwer, Dordrecht, The Netherlands, pp. 801–818.

Ainslie, G. (1975) Specious reward: a behavioural theory of impulsiveness and impulse control. *Psychological Bulletin* 82, 463–496.

Alvey, D.M. (1989) Evaluation of two perchery designs for egg production. *British Poultry Science* 30, 960.

American Poultry Association (1989) *American Standard of Perfection.* American Poultry Association, Estacada, Oregon.

Amgarten, M. and Mettler, A. (1989) Economical consequences of the introduction of alternative housing systems for laying hens in Switzerland. In: Faure, J.M. and Mills, A.D. (eds)

Proceedings, Third European Symposium on Poultry Welfare. World's Poultry Science Association, Tours, France, pp. 213–228.

Anderson, D.P., Beard, C.W. and Hanson, R.P. (1964) Adverse effects of ammonia on the surface ultrastructure of the lung and trachea of broiler chickens. *Poultry Science* 64, 2056–2061.

Animal Welfare Institute (1991) *Animals and Their Legal Rights. A Survey of American Laws from 1641 to 1991*. Animal Welfare Institute, Washington, DC.

Anjum, A.D., Payne, L.N. and Appleby, E.C. (1989) Oviduct magnum tumours in the domestic fowl and their association with laying. *Veterinary Record* 125, 42–43.

Anonymous (1983) Perchery tries again to match cages. *Poultry World* 12th August, 10.

Anonymous (1989) *How Astrid Lindgren Achieved Enactment of the 1988 Law Protecting Farm Animals in Sweden*. Animal Welfare Institute, Washington, DC.

Anonymous (2001) Disruptions continue for Arkansas poultry. *Feedstuffs* 15th January, 6.

Appleby, M.C. (1984) Factors affecting floor laying by domestic hens: a review. *World's Poultry Science Journal* 40, 241–249.

Appleby, M.C. (1985) Hawks, doves . . . and chickens. *New Scientist* 1438, 16–18.

Appleby, M.C. (1990) Behaviour of laying hens in cages with nest sites. *British Poultry Science* 31, 71–81.

Appleby, M.C. (1993) Should cages for laying hens be banned or modified? *Animal Welfare* 2, 67–80.

Appleby, M.C. (1995) Perch length in cages for medium hybrid hens. *British Poultry Science* 36, 23–31.

Appleby, M.C. (1998) The Edinburgh Modified Cage: effects of group size and space allowance on brown laying hens. *Journal of Applied Poultry Research* 7, 152–161.

Appleby, M.C. (1999) *What Should We Do About Animal Welfare?* Blackwell Science, Oxford.

Appleby, M.C. (2004) What causes crowding? Effects of space, facilities and group size on behaviour, with particular reference to furnished cages for hens. *Animal Welfare* 13 (in press).

Appleby, M.C. and Duncan, I.J.H. (1989) Development of perching in hens. *Biology of Behaviour* 14, 157–168.

Appleby, M.C. and Hughes, B.O. (1990) Cages modified with perches and nest sites for the improvement of welfare. *World's Poultry Science Journal* 46, 38–40.

Appleby, M.C. and Hughes, B.O. (1991) Welfare of laying hens in cages and alternative systems: environmental, physical and behavioural aspects. *World's Poultry Science Journal* 47, 109–126.

Appleby, M.C. and Lawrence, A.B. (1987) Food restriction as a cause of stereotypic behaviour in tethered gilts. *Animal Production* 45, 103–110.

Appleby, M.C. and McRae, H.E. (1986) The individual nest box as a superstimulus for domestic hens. *Applied Animal Behaviour Science* 15, 169–176.

Appleby, M.C. and Sandøe, P. (2002) Philosophical debates relevant to animal welfare: the nature of well-being. *Animal Welfare* 11, 283–294.

Appleby, M.C. and Waran, N.K. (1997) Physical conditions. In: Appleby, M.C. and Hughes, B.O. (eds) *Animal Welfare*. CAB International, Wallingford, UK, pp. 177–190.

Appleby, M.C., McRae, H.E. and Duncan, I.J.H. (1983) Nesting and floor laying by domestic hens: effects of individual variation in perching behaviour. *Behaviour Analysis Letters* 3, 345–352.

Appleby, M.C., McRae, H.E., Duncan, I.J.H. and Bisazza, A. (1984a) Choice of social conditions by laying hens. *British Poultry Science* 25, 111–117.

Appleby, M.C., McRae, H.E. and Peitz, B.E. (1984b) The effect of light on the choice of nests by domestic hens. *Applied Animal Ethology* 11, 249–254.

Appleby, M.C., Maguire, S.N. and McRae, H.E. (1985) Movement by domestic fowl in commercial flocks. *Poultry Science* 64, 48–50.

Appleby, M.C., Maguire, S.N. and McRae, H.E. (1986) Nesting and floor laying by domestic hens in a commercial flock. *British Poultry Science* 27, 75–82.

Appleby, M.C., Duncan, I.J.H. and McRae, H.E. (1988a) Perching and floor laying by domestic hens: experimental results and their commercial application. *British Poultry Science* 29, 351–357.

Appleby, M.C., Hogarth, G.S., Anderson, J.A., Hughes, B.O. and Whittemore, C.T. (1988b) Performance of a deep litter system for egg production. *British Poultry Science* 29, 735–751.

Appleby, M.C., Hogarth, G.S. and Hughes, B.O. (1988c) Nest box design and nesting material in a deep litter house for laying hens. *British Poultry Science* 29, 215–222.

Appleby, M.C., Hughes, B.O. and Hogarth, G.S. (1989) Behaviour of laying hens in a deep litter house. *British Poultry Science* 30, 545–553.

Appleby, M.C., Hughes, B.O. and Elson, H.A. (1992) *Poultry Production Systems: Behaviour, Management and Welfare.* CAB International, Wallingford, UK.

Appleby, M.C., Smith, S.F. and Hughes, B.O. (1993) Nesting, dust bathing and perching by laying hens in cages: effects of design on behaviour and welfare. *British Poultry Science* 34, 835–847.

Appleby, M.C., Hughes, B.O. and Savory, C.J. (1994) Current state of poultry welfare: progress, problems and strategies. *British Poultry Science* 35, 467–475.

Appleby, M.C., Walker, A.W., Nicol, C.J., Lindberg, A.C., Freire, R., Hughes, B.O. and Elson, H.A. (2002) Development of furnished cages for laying hens. *British Poultry Science* 43, 489–500.

Ashcoff, J. and Meyer-Lohmann, J. (1954) Angeborene 24-Stunden-Periodik beim Kuken. *Pflugers Archiv fur die gesamte Physiologie des Menschen und die Tiere* 260, 170–176.

Bang, B.G. and Wenzel, B.M. (1985) Nasal cavity and olfactory system. In: King, A.S. and McLelland, J. (eds) *Form and Function in Birds*, Vol. 3. Academic Press, London, pp. 195–225.

Barnard, C.J. and Burk, T. (1979) Dominance hierarchies and the evolution of 'individual recognition'. *Journal of Theoretical Biology* 81, 65–73.

Barnard, C.J. and Hurst, J.L. (1996) Welfare by design: the natural selection of welfare criteria. *Animal Welfare* 5, 405–433.

Baxter, M.R. (1983) Ethology in environmental design. *Applied Animal Ethology* 9, 207–220.

Bekoff, M. (1998) *Animal Play: Evolutionary, Comparative and Ecological Perspectives.* Cambridge University Press, Cambridge.

Bell, D.D. (2002) Introduction to the table-egg industry. In: Bell, D.D. and Weaver, W.D. (eds) *Commercial Chicken Meat and Egg Production*, 5th edn. Kluwer, Dordrecht, The Netherlands, pp. 945–963.

Bell, D.D. and Swanson, M.H. (1975) *Crowding of Chickens in Cages Reduces your Profits.* Leaflet 2273, University of California, California.

Bell, D.J. and Freeman, B.M. (1971) *Physiology and Biochemistry of the Domestic Fowl*, Vols 1–3. Academic Press, London.

Belshaw, R.H.H. (1985) *Guinea Fowl of the World.* Ninitod Book Services, Liss, UK.

Bennett, R.M. (1997) Economics. In: Appleby, M.C. and Hughes, B.O. (eds) *Animal Welfare.* CAB International, Wallingford, UK, pp. 235–248.

Bentham, J. (1789) *Introduction to the Principles of Morals and Legislation.* 1996 Imprint. Clarendon Press, Oxford.

Bermond, B. (1997) The myth of animal suffering. In: Dol, M., Kasanmoentalib, S., Lijmbach, S., Rivas, E. and van den Bos, R. (eds) *Animal Consciousness and Animal Ethics.* Van Gorcum, Assen, The Netherlands, pp. 125–143.

Bernon, D.B. and Siegel, P.B. (1983) Mating frequency in male chickens: crosses among selected and unselected lines. *Genetic Selection and Evolution* 15, 445–454.

Bertram, B.C.R. (1992) *The Ostrich Communal Nesting System*. Princeton University Press, Princeton, New Jersey.

Bessei, W. (1973) Die selective Futteraufnahme beim Huhn. *Deutsche Geflugelwirtschaft and Schweineproducktion* 25, 107–109.

Bessei, W. (1986) Das Verhalten des Huhns in der Intensivhaltung. *Joahrbuch der Geflugel Produktion* 95–99.

Bhagwat, A.L. and Craig, J.V. (1975) Fertility from natural matings influenced by social and physical environments in multiple-bird cages. *Poultry Science* 54, 222–227.

Biggs, P.M. (1990) Vaccines and vaccination – past, present and future. *British Poultry Science* 31, 3–22.

Bilĉík, B. and Keeling, L.J. (1999) Changes in feather condition in relation to feather pecking and aggressive behaviour in laying hens. *British Poultry Science* 40, 444–451.

Blokhuis, H.J. (1983) The relevance of sleep in poultry. *World's Poultry Science Journal* 39, 33–37.

Blokhuis, H.J. (1989) The effect of a sudden change in floor type on pecking behaviour in chicks. *Applied Animal Behaviour Science* 22, 65–73.

Blokhuis, H.J. and van der Haar, J.W. (1989) Effects of floor type during rearing and of beak trimming on ground pecking and feather pecking in laying hens. *Applied Animal Behaviour Science* 22, 359–369.

Bogner, H., Pescke, W., Seda, V. and Popp, K. (1979) Studie zum Flaschenbedarf von Legehennen in Kafigen bei Bestimmen Activitaten. *Berliner und Munchener Tierzartliche Wochenschrift* 92, 340–343.

Borchelt, P.L. and Overmann, S.R. (1974) Development of dustbathing in Bobwhite quail. I. Effects of age, experience, texture of dust, strain, and social facilitation. *Developmental Psychobiology* 7, 305–313.

Braastad, B.O. (1990) Effects on behaviour and plumage of a key-stimuli floor and a perch in triple cages for laying hens. *Applied Animal Behaviour Science* 27, 127–139.

Braastad, B.O. and Katle, J. (1989) Behavioural differences between laying hen populations selected for high and low efficiency of food utilisation. *British Poultry Science* 30, 533–544.

Brake, J. (1987) Influence of presence of perches during rearing on incidence of floor laying in broiler breeders. *Poultry Science* 66, 1587–1589.

Brant, A.W. (1998) A brief history of the turkey. *World's Poultry Science Journal* 54, 365–373.

Bremond, J.C. (1963) Acoustic behavior of birds. In: Busnel, R.G. (ed.) *Acoustic Behavior of Animals*. Elsevier, London.

Brillard, J.P. and McDaniel, G.R. (1986) Influence of spermatozoa numbers and insemination frequency on fertility in dwarf broiler breeder hens. *Poultry Science* 65, 2330–2334.

Brillard, J.P., Galut, O. and Nys, Y. (1987) Possible causes of subfertility in hens following insemination near the time of oviposition. *British Poultry Science* 28, 307–318.

British Egg Information Service (1990) *Egg Farming*. British Egg Information Service, London.

Brocklehurst, D.S. (1975) *A Preliminary Report on a Survey of Floor Laying in Breeding Stock*. East of Scotland College of Agriculture, Edinburgh, UK.

Broom, D.M. (1990) Effects of handling and transport on laying hens. *World's Poultry Science Journal* 46, 48–50.

Broom, D.M. and Johnson, K.G. (1993) *Stress and Animal Welfare*. Chapman and Hall, London.

Brown, E. (1929) *Poultry Breeding and Production*, Vol. 1. Ernest Benn, London.

Bubier, N.E., Paxton, C.G.M., Bowers, P. and Deeming, D.C. (1998) Courtship behaviour of farmed ostriches in relation to humans. *British Poultry Science* 39, 477–481.

Buchholz, R. (1995) Female choice, parasite load and male ornamentation in wild turkeys. *Animal Behaviour* 50, 929–943.

Calet, C. (1965) The relative value of pellets versus mash and grain in poultry nutrition. *World's Poultry Science Journal* 21, 23–52.

Candland, D.K., Taylor, D.B., Dresdale, L., Leiphart, J.M. and Solow, S.P. (1969) Heart rate, aggression and dominance in the domestic chicken. *Journal of Comparative and Physiological Psychology* 69, 70–76.

Carpenter, G.A., Smith, W.K., MacLaren, A.P.C. and Spackman, D. (1986) Effect of internal air filtration on the performance of broilers and the aerial concentrations of dust and bacteria. *British Poultry Science* 27, 471–480.

Carruthers, P. (1992) *The Animals Issue: Moral Theory in Practice.* Cambridge University Press, Cambridge.

Carter, T.C. (1971) The hen's egg: shell cracking at oviposition and its inheritance. *British Poultry Science* 12, 259–278.

Centrum voor Onderzoek en Voorlichting voor der Pluimveehouderij (1988) *The Tiered Wire Floor System for Laying Hens – Development and Testing of an Alternative Aviary for Laying Hens, 1980–1987.* COVP Spelderholt Report No. 484, Beekbergen, The Netherlands.

Cherry, P. and Barwick, M.W. (1962) The effect of light on broiler growth. 1. Light intensity and colour. *British Poultry Science* 3, 31–39.

Christensen, V.L. and Bagley, L.G. (1989) Efficacy of fertilization in artificially inseminated turkey hens. *Poultry Science* 68, 724–729.

Classen, H.L. and Stevens, J.P. (1995) Nutrition and growth. In: Hunton, P. (ed.) *Poultry Production.* Elsevier, Amsterdam, pp. 79–99.

Clayton, G.A., Lake, P.E., Nixey, C., Jones, D.R., Charles, D.R., Hopkins, J.R., Binstead, J.A. and Pickett, R. (1985) *Turkey Production: Breeding and Husbandry.* Ministry of Agriculture, Fisheries and Food reference book 242. HMSO, London.

Cloutier, S., Newberry, R.C., Honda, K. and Alldredge, R. (2002) Cannibalistic behaviour spread by social learning. *Animal Behaviour* 63, 1153–1162.

Cochlan, A. (1993) Pressure group broods over altered turkeys. *New Scientist* 29th April, 9.

Coenen, A.M.L., Wolters, E.M.T.J., van Luijtelaar, E.L.J.M. and Blokhuis, H.J. (1988) Effects of intermittent lighting on sleep and activity in the domestic hen. *Applied Animal Behaviour Science* 20, 309–318.

Collias, N.E. (1952) The development of social behavior in birds. *Auk* 69, 127–159.

Collias, N.E. (1962) The behaviour of ducks. In: Hafez, E.S.E. (ed.) *The Behaviour of Domestic Animals.* Ballière, Tindall and Cox, London, pp. 565–585.

Collias, N.E. (1987) The vocal repertoire of the red junglefowl: a spectrographic classification and the code of communication. *Condor* 89, 510–524.

Collias, N.E. (2000) Filial imprinting and leadership among chicks in family integration of the domestic fowl. *Behaviour* 137, 197–211.

Collias, N.E. and Collias, E.C. (1967) A field study of the Red Jungle Fowl in North-Central India. *Condor* 69, 360–386.

Collias, N.E. and Collias, E.C. (1996) Social organization of red junglefowl, *Gallus gallus,* population related to evolution theory. *Animal Behaviour* 51, 1337–1354.

Collias, N.E., Collias, E.C., Hunsaker, D. and Mining, L. (1966) Locality fixation, mobility and social organization within an unconfined population of Red Jungle Fowl. *Animal Behaviour* 14, 550–559.

Coltherd, J.B. (1966) The domestic fowl in Ancient Egypt. *Ibis* 108, 217–223.

Commission of the European Communities (1985) Amendment 1943/85 to Regulation 95/69, also amended by 927/69 and 2502171. *Official Journal of the European Communities* 13th July.

Commission of the European Communities (1986) Council Directive 86/113/EEC: Welfare of Battery Hens. *Official Journal of the European Communities* (L 95) 29, 45–49.

Commission of the European Communities (1999) Council Directive 99/74/EC of 19 July 1999: Laying Down Minimum Standards for the Protection of Laying Hens. *Official Journal of the European Communities* L 203–207.

Cooper, D.M. (1969) The use of artificial insemination in poultry breeding, the evaluation of semen dilution and storage. In: Carter, T.C. and Freeman, B.M. (eds) *The Fertility and Hatchability of the Hen's Egg.* Oliver and Boyd, Edinburgh, UK, pp. 31–44.

Cooper, J.J. and Appleby, M.C. (1995) Nesting behaviour of hens: effects of experience on motivation. *Applied Animal Behaviour Science* 42, 283–295.

Cooper, J.J. and Appleby, M.C. (2003) The value of environmental resources to domestic hens: a comparison of the work-rate for food and for nests as a function of time. *Animal Welfare* 12, 39–52.

Cornetto, T. and Estevez, I. (2001) Behavior of the domestic fowl in the presence of vertical panels. *Poultry Science* 80, 1455–1465.

Cowan, P.J., Michie, W. and Roele, D.J. (1978) Choice feeding of the egg type pullet. *British Poultry Science* 19, 153–157.

Craig, J.V. (1980) Training colony-cage pullets to use nests in mating pens. *Poultry Science* 59, 1596.

Craig, J.V. (1981) *Domestic Animal Behavior: Causes and Implications for Animal Care and Management.* Prentice Hall, Englewood Cliffs, New Jersey.

Craig, J.V. and Bhagwat, A.L. (1974) Agonistic and mating behavior of adult chickens modified by social and physical environments. *Applied Animal Ethology* 1, 57–65.

Craig, J.V. and Muir, W.M. (1993) Selection for reduced beak-inflicted injuries among caged hens. *Poultry Science* 72, 411–420.

Craig, J.V., Biswas, D.K. and Guhl, A.M. (1969) Agonistic behaviour influenced by strangeness, crowding and heredity in female domestic fowl (*Gallus domesticus*). *Animal Behaviour* 17, 498–506.

Craig, J.V., Al-Rawi, B.A. and Kratzer, D.D. (1977) Social status and sex ratio effects on mating frequency of cockerels. *Poultry Science* 56, 762–772.

Crawford, R.D. (1990) Origin and history of poultry species. In: Crawford, R.D. (ed.) *Poultry Breeding and Genetics.* Elsevier, Amsterdam, pp. 1–41.

Cruickshank, G. (1999) Egg producers call for unity to fight cage ban. *Poultry World* August, 1–3.

Cunningham, D.L. and van Tienhoven, A. (1983) Relationship between production factors and dominance in White Leghorn hens in a study on social rank and cage design. *Applied Animal Ethology* 11, 33–44.

Curtis, P.E. (1990) *A Handbook of Poultry and Game Bird Diseases*, 3rd edn. Liverpool University Press, Liverpool, UK.

Curtis, S.E. (1983) *Environmental Management in Animal Agriculture.* Iowa State University Press, Ames, Iowa.

Dantzer, R. (1986) Behavioral, physiological and functional aspects of stereotyped behavior: a review and a re-interpretation. *Journal of Animal Science* 62, 1776–1786.

Davison, T.F. (2003) The immunologist's debt to the chicken. *British Poultry Science* 44, 6–21.

Dawkins, M.S. (1981) Priorities in the cage size and flooring preferences of domestic hens. *British Poultry Science* 22, 255–263.

Dawkins, M.S. (1985) Cage height preference and use in battery-kept hens. *Veterinary Record* 116, 345–347.

Dawkins, M.S. (1988) Behavioural deprivation: a central problem in animal welfare. *Applied Animal Behaviour Science* 20, 209–225.

Dawkins, M.S. (1989) Time budgets in Red Jungle Fowl as a basis for the assessment of welfare in domestic fowl. *Applied Animal Behaviour Science* 24, 70–77.

Dawkins, M.S. (1990) From an animal's point of view: motivation, fitness and animal welfare. *Behavioral and Brain Sciences* 13, 1–9.

Dawkins, M.S. and Beardsley, T. (1986) Reinforcing properties of access to litter in hens. *Applied Animal Behaviour Science* 15, 351–364.

Dawkins, M.S. and Hardie, S. (1989) Space needs of laying hens. *British Poultry Science* 30, 413–416.

Dawkins, M.S. and Nicol, C.J. (1989) No room for manoeuvre. *New Scientist* 16th September, 44–46.

Dawkins, M.S., Cook, P.A., Whittingham, M.J., Mansell, K.A. and Harper, A.E. (2003) What makes free-range broiler chickens range? *In situ* measurement of habitat preference. *Animal Behaviour* 66, 151–160.

Deaton, J.W., Lott, B.D., Branton, S.L. and Simmons, J.D. (1987) Research note: effect of beak trimming on body weight and feed intake of egg-type pullets fed pellets or mash. *Poultry Science* 66, 1552–1554.

Decuypere, E., Buyse, J. and Buys, N. (2000) Ascites in broiler chickens: exogenous and endogenous structural and functional causal factors. *World's Poultry Science Journal* 56, 367–377.

Deeming, D.C. (ed.) (1999a) *The Ostrich: Biology, Production and Health*. CAB International, Wallingford, UK.

Deeming, D.C. (1999b) Introduction. In: Deeming, D.C. (ed.) *The Ostrich: Biology, Production and Health*. CAB International, Wallingford, UK, pp. 1–11.

Deeming, D.C. and Angel, C.R. (1996) Introduction to the ratites and farming operations around the world. In: *Improving our Understanding of Ratites in a Farming Environment*. Ratite Conference, Oxfordshire, UK, pp. 1–4.

Deeming, D.C. and Ar, A. (1999) Factors affecting the success of commercial incubation. In: Deeming, D.C. (ed.) *The Ostrich: Biology, Production and Health*. CAB International, Wallingford, UK, pp. 159–190.

Deeming, D.C. and Bubier, N.E. (1999) Behaviour in natural and captive environments. In: Deeming, D.C. (ed.) *The Ostrich: Biology, Production and Health*. CAB International, Wallingford, UK, pp. 83–104.

Department of Environment, Food and Rural Affairs (2000) *Agricultural Statistics for the United Kingdom*. DEFRA Publications, London.

Department of Environment, Food and Rural Affairs (2002a) *Code of Recommendations for the Welfare of Livestock: Laying Hens*. DEFRA Publications, London.

Department of Environment, Food and Rural Affairs (2002b) *Code of Recommendations for the Welfare of Livestock: Meat Chickens and Breeding Chickens*. DEFRA Publications, London.

Desforges, M.F. and Wood-Gush, D.G.M. (1975a) A behavioural comparison of mallard and domestic duck. Habituation and flight reactions. *Animal Behaviour* 23, 692–697.

Desforges, M.F. and Wood-Gush, D.G.M. (1975b) A behavioural comparison of mallard and domestic duck. Spatial relations in small flocks. *Animal Behaviour* 23, 698–705.

Dingle, J. (1999) Case studies. In: *Ostrich Production Systems. FAO Paper 144*. Food and Agriculture Organization, Rome, Italy.

Domjan, M. and Nash, S. (1988) Stimulus control of social behaviour in male Japanese quail, *Coturnix coturnix japonica*. *Animal Behaviour* 36, 1006–1015.

Dorminey, R.M. (1974) Incidence of floor eggs as influenced by time of nest installation, artificial lighting and nest location. *Poultry Science* 53, 1886–1891.

Druce, C. (1993) The stress of artificial insemination in turkeys. In: Savory, C.J. and Hughes, B.O. (eds) *Proceedings, Fourth European Symposium on Poultry Welfare*. Universities Federation for Animal Welfare, Potters Bar, UK, p. 281.

Duff, S.R.I., Hocking, P.M., Randall, C.J. and MacKenzie, G. (1989) Head swelling of traumatic aetiology in broiler breeder fowl. *Veterinary Record* 126, 133–134.

Duncan, E.T., Appleby, M.C. and Hughes, B.O. (1992) Effect of perches in laying cages on welfare and production of hens. *British Poultry Science* 33, 25–35.

Duncan, I.J.H. (1970) Frustration in the fowl. In: Freeman, B.M. and Gordon, R.F. (eds) *Aspects of Poultry Behaviour*. British Poultry Science, Edinburgh, UK, pp. 15–31.

Duncan, I.J.H. (1978a) An overall assessment of poultry welfare. In: Sorensen, L.Y. (ed.) *Proceedings, First Danish Seminar on Poultry Welfare*. World's Poultry Science Association, Copenhagen, Denmark.

Duncan, I.J.H. (1978b) The interpretation of preference tests in animal behaviour. *Applied Animal Ethology* 4, 197–200.

Duncan, I.J.H. (1982) Investigations into the feelings of the domestic fowl: what's all the flap about? *CEC Research on Poultry Welfare. Progress Reports 1981/1982*. Commission of the European Communities, Brussels, Belgium.

Duncan, I.J.H. (1996) Animal welfare defined in terms of feelings. *Acta Agriculturae Scandinavica, Supplementum* 27, 29–35.

Duncan, I.J.H. and Fraser, D. (1997) Understanding animal welfare. In: Appleby, M.C. and Hughes, B.O. (eds) *Animal Welfare*. CAB International, Wallingford, UK, pp. 19–31.

Duncan, I.J.H. and Hughes, B.O. (1972) Free and operant feeding in domestic fowls. *Animal Behaviour* 20, 775–777.

Duncan, I.J.H. and Hughes, B.O. (1988) Can the welfare needs of poultry be measured? In: Hardcastle, J. (ed.) *Science and the Poultry Industry*. Agricultural and Food Research Council, London, pp. 24–25.

Duncan, I.J.H. and Kite, V. (1987) Some investigations into motivation in the domestic fowl. *Applied Animal Behaviour Science* 18, 387–388.

Duncan, I.J.H. and Kite, V. (1989) Nest site selection and nest building behaviour in domestic fowl. *Animal Behaviour* 37, 215–231.

Duncan, I.J.H. and Mench, J.A. (1993) Behaviour as an indicator of welfare in various systems. In: Savory, C.J. and Hughes, B.O. (eds) *Proceedings, Fourth European Symposium on Poultry Welfare*. Universities Federation for Animal Welfare, Potters Bar, UK, pp. 69–80.

Duncan, I.J.H. and Wood-Gush, D.G.M. (1972) An anlaysis of displacement preening in the domestic fowl. *Animal Behaviour* 20, 68–71.

Duncan, I.J.H., Savory, C.J. and Wood-Gush, D.G.M. (1978) Observations on the reproductive behaviour of domestic fowl in the wild. *Applied Animal Ethology* 4, 29–42.

Duncan, I.J.H., Slee, G.S., Kettlewell, P., Berry, P. and Carlisle, A.J. (1986) Comparison of the stressfulness of harvesting broiler chickens by machine and by hand. *British Poultry Science* 27, 109–114.

Duncan, I.J.H., Hocking, P.M. and Seawright, E. (1990) Sexual behaviour and fertility in broiler breeder domestic fowl. *Applied Animal Behaviour Science* 26, 201–213.

Duncan, I.J.H., Widowski, T.M., Malleau, A.E., Lindberg, A.C. and Petherick, J.C. (1998) External factors and causation of dustbathing in domestic hens. *Behavioural Processes* 43, 219–228.

Ede, D.A. (1964) *Bird Structure: an Approach through Evolution, Development and Function in the Fowl*. Hutchinson, London.

Ehlhardt, D.A. and Koolstra, C.L.M. (1984) Multi-tier system for housing laying hens. *Pluimveehouderij* 21st December, 44–47.

Ekstrand, C. (1998) An observational cohort study on the effects of catching methods on carcase rejection rates in broilers. *Animal Welfare* 7, 87–96.

Elson, H.A. (1981) Modified cages for layers. In: Universities Federation for Animal Welfare (ed.) *Alternatives to Intensive Husbandry Systems*. Universities Federation for Animal Welfare, Potters Bar, UK, pp. 47–50.

Elson, H.A. (1985) The economics of poultry welfare. In: Wegner, R.M. (ed.) *Proceedings,*

Second European Symposium on Poultry Welfare. World's Poultry Science Association, Celle, Germany, pp. 244–253.

Elson, H.A. (1988) Making the best cage decisions. In: Cambridge Poultry Conference (ed.) *Cages for the Future.* Agricultural Development and Advisory Service, Nottingham, UK, pp. 70–76.

Emmans, G.C. (1975) Problems in feeding laying hens: can a system based on choice solve them? *World's Poultry Science Journal* 31, 31.

Emmans, G.C. (1977) The nutrient intake of laying hens given a choice of diets in relation to their production requirements. *British Poultry Science* 18, 227–236.

Emmans, G.C. and Charles, D.R. (1977) Climatic environment and poultry feeding in practice. In: Haresign, W., Swan, H. and Lewis, D. (eds) *Nutrition and the Climatic Environment.* Butterworths, London, pp. 31–49.

England, C. (2002) Burger King and animal welfare: why did this company get involved? In: *Proceedings, Canadian Association for Laboratory Animal Science and Alberta Farm Animal Care Conference.* Canadian Association for Laboratory Animal Science, Edmonton, Canada, p. 13.

Englemann, C. (1940) Versuche uber die 'Beliebtheit' eineger Getreidearten beim Huhn. *Zeitschrift fur Vergleichende Physiologie* 27, 526–634.

Esmay, M.L. (1978) *Principles of Animal Environment.* AVI, Westport, Connecticut.

Estevez, I., Newberry, R.C. and de Reyna, L.A. (1997) Broiler chickens: a tolerant social system? *Etologia* 5, 19–29.

Etches, R.J. (1996) *Reproduction in Poultry.* CAB International, Wallingford, UK.

European Communities (2000) European Communities proposal: animal welfare and trade in agriculture. *WTO Committee on Agriculture, Special Session, Paper G/AG/NG/W/19.* Available at: www.wto.org (Trade Topics, Agriculture, Negotiations)

Evans, R.M. (1975) Stimulus intensity and acoustical communication in young domestic chicks. *Behaviour* 55, 73–80.

Ewbank, R., Kim-Madslien, F. and Hart, C.B. (eds) (1999) *Management and Welfare of Farm Animals: the UFAW Handbook,* 4th edn. Universities Federation for Animal Welfare, Potters Bar, UK.

Farm Animal Welfare Council (1986) *An Assessment of Egg Production Systems.* FAWC, Tolworth, UK.

Farm Animal Welfare Council (1991) *Report on the Welfare of Laying Hens in Colony Systems.* FAWC, Tolworth, UK.

Farm Animal Welfare Council (1997) *Report on the Welfare of Laying Hens.* FAWC, Tolworth, UK.

Farrant, J. (1999) IEC's world action to keep cages. *Poultry World* November, 1–4.

Farsaie, A., Carr, L.E. and Wabeck, C.J. (1983) Mechanical harvest of broilers. *Transactions ASAE* 26, 650–653.

Faure, J.M. and Jones, R.B. (1982) Effects of age, access and time of day on perching behaviour in the domestic fowl. *Applied Animal Ethology* 8, 357–364.

Faure, J.M. and Lagadic, H. (1994) Elasticity of demand for food and sand in laying hens subjected to variable wind speed. *Applied Animal Behaviour Science* 42, 49–59.

Fischer, G.J. (1975) The behaviour of chickens. In: Hafez, E.S.E. (ed.) *The Behaviour of Domestic Animals,* 3rd edn. Ballière Tindall, London, pp. 454–489.

Fisher, C. and Bowles, D. (2002) *Hard-boiled Reality: Animal Welfare-friendly Egg Production in a Global Market.* Royal Society for the Protection of Animals, Horsham, UK.

Food and Agriculture Organization (1997) *Production Yearbook.* FAO, United Nations, New York.

Frankenhuis, M.T., Vertommen, M.H. and Hemminga, H. (1991) Influence of claw clipping, stocking density and feeding space on the incidence of scabby hips in broilers. *British Poultry Science* 32, 227–230.

Fraser, A.F. (1992) *The Behaviour of the Horse*. CAB International, Wallingford, UK.

Fraser, D. and Duncan, I.J.H. (1998) 'Pleasures', 'pains' and animal welfare: toward a natural history of affect. *Animal Welfare* 7, 383–396.

Fraser, D. and Matthews, L.R. (1997) Preference and motivation testing. In: Appleby, M.C. and Hughes, B.O. (eds) *Animal Welfare*. CAB International, Wallingford, UK, pp. 159–173.

Fraser, D., Weary, D.M., Pajor, E.A. and Milligan, B.N. (1997) A scientific conception of animal welfare that reflects ethical concerns. *Animal Welfare* 6, 187–205.

Freeman, B.M. (1983) Floor space allowance for caged domestic fowl. *Veterinary Record* 112, 562–563.

Freeman, B.M. (1983–1984) *Physiology and Biochemistry of the Domestic Fowl*, Vols 4 and 5. Academic Press, London.

Freire, R., Appleby, M.C. and Hughes, B.O. (1997) The interaction between pre-laying behaviour and feeding in hens: implications for motivation. *Behaviour* 134, 1019–1030.

Freire, R., Appleby, M.C. and Hughes, B.O. (1998) Effects of social interactions on pre-laying behaviour in hens. *Applied Animal Behaviour Science* 56, 47–57.

Fritz, J., Bisenberger, A. and Kortrschal K. (2000) Stimulus enhancement in greylag geese: socially mediated learning of an operant task. *Animal Behaviour* 59, 1119–1125.

Fröhlich, E.K.F. and Oester, H. (2001) From battery cages to aviaries: 20 years of Swiss experience. In: Oester, H. and Wyss, C. (eds) *Proceedings, Sixth European Symposium on Poultry Welfare*. World's Poultry Science Association, Zollikofen, Switzerland, pp. 51–59.

Fujita, H. (1973) Quantitative studies on the variations in feeding activity of chickens. II. Effect of the physical form of the feed on the feeding activity of laying hens. *Japanese Poultry Science* 10, 47–54.

Furlow, B., Kimball, R.T. and Marshall, M.C. (1998) Are rooster crows honest signals of fighting ability? *Auk* 115, 763–766.

Gallagher, J.E. (1977) Sexual imprinting: a sensitive period in Japanese quail (*Coturnix coturnix japonica*). *Animal Learning and Behavior* 6, 363–365.

Garner, R. (1998) *Political Animals: Animal Protection Politics in Britain and the United States*. Macmillan, London.

Gentle, M.J. (1975) Gustatory behaviour of the chicken and other birds. In: Wright, P., Caryl, P.G. and Vowles, D.M. (eds) *Neural and Endocrine Aspects of Behaviour in Birds*. Elsevier, Amsterdam, pp. 305–318.

Gentle, M.J. (1986a) Aetiology of food-related oral lesions in chickens. *Research in Veterinary Science* 40, 219–224.

Gentle, M.J. (1986b) Beak trimming in poultry. *World's Poultry Science Journal* 42, 268–275.

Gentle, M.J. (1997) Acute and chronic pain in the chicken. In: Koene, P. and Blokhuis, H.J. (eds) *Proceedings, Fifth European Symposium on Poultry Welfare*. Wageningen Agricultural University, Wageningen, The Netherlands, pp. 5–11.

Gentle, M.J. and Hunter, L.N. (1991) Physiological and behavioural responses associated with feather removal in *Gallus gallus* var *domesticus*. *Research in Veterinary Science* 50, 95–101.

Gentle, M.J., Hughes, B.O. and Hubrecht, R.C. (1982) The effect of beak trimming on food intake, feeding behaviour and body weight in adult hens. *Applied Animal Ethology* 8, 147–159.

Gentle, M.J., Hughes, B.O., Fox, A. and Waddington, D. (1997) Behavioural and anatomical consequences of two beak trimming methods in 1- and 10-d-old domestic chicks. *British Poultry Science* 38, 453–463.

Gerken, M. and Mills, A.D. (1993) Welfare of domestic quail. In: Savory, C.J. and Hughes, B.O. (eds) *Proceedings, Fourth European Symposium on Poultry Welfare*. Universities Federation for Animal Welfare, Potters Bar, UK, pp. 158–176.

Gibson, S.W., Innes, J. and Hughes, B.O. (1985) Aggregation behaviour of laying fowls in a covered strawyard. In: Wegner, R.M. (ed.) *Proceedings, Second European Symposium on Poultry Welfare.* World's Poultry Science Association, Celle, Germany, pp. 295–298.

Gibson, S.W., Dun, P. and Hughes, B.O. (1988) The performance and behaviour of laying fowls in a covered strawyard system. *Research and Development in Agriculture* 5, 153–163.

Goldstein, D.L. and Skadhauge, E. (2000) Renal and extrarenal regulation of body fluid composition. In: Whittow, G.C. (ed.) *Sturkie's Avian Physiology*, 5th edn. Academic Press, San Diego, California, pp. 265–297.

Gonyou, H.W. (1993) Animal welfare: definitions and assessment. *Journal of Agricultural and Environmental Ethics* 6, Supplement 2, 37–43.

Gooderham, K.R. (1993) Disease prevention and control in ducks. In: Pattison, M. (ed.) *The Health of Poultry.* Longman, Harlow, UK, pp. 239–264.

Graves, H.B., Hable, C.P. and Jenkins, T.H. (1985) Sexual selection in *Gallus*: effects of morphology and dominance on female spatial behavior. *Behavioural Processes* 11, 189–197.

Gray, P.H. and Howard, K.I. (1957) Specific recognition of humans in imprinted chicks. *Perceptual and Motor Skills* 7, 301–304.

Green, L.E., Lewis, K., Kimpton, A. and Nicol, C.J. (2000) Cross-sectional study of the prevalence of feather pecking in laying hens in alternative systems and its associations with management and disease. *Veterinary Record* 147, 233–238.

Gregory, N.G. and Wilkins, L.J. (1989) Broken bones in domestic fowl: handling and processing damage in end-of-lay battery hens. *British Poultry Science* 30, 555–562.

Gregory, N.G., Wilkins, L.J., Eleperuma, S.D., Ballantyne, A.J. and Overfield, N.D. (1990) Broken bones in domestic fowls: effects of husbandry system and stunning method in end-of-lay hens. *British Poultry Science* 31, 59–69.

Griffin, H.D. and Goddard, C. (1994) Rapidly growing broiler (meat-type) chickens – their origin and use for comparative studies in the regulation of growth. *International Journal of Biochemistry* 26, 19–28.

Grigor, P.N., Hughes, B.O. and Appleby, M.C. (1995a) Effects of regular handling and exposure to an outside area on subsequent fearfulness and dispersal in domestic hens. *Applied Animal Behaviour Science* 44, 47–55.

Grigor, P.N., Hughes, B.O. and Appleby, M.C. (1995b) Emergence and dispersal behaviour in domestic hens: effects of social rank and novelty of an outdoor area. *Applied Animal Behaviour Science* 45, 97–108.

Grubb, B.R. (1983) Allometric relations of cardiovascular function in birds. *American Journal of Physiology* 245, H567–H562.

Guhl, A.M. (1949) Heterosexual dominance and mating behaviour in chickens. *Behaviour* 2, 106–120.

Guhl, A.M (1953) Social behavior of the domestic fowl. *Technical Bulletin of the Kansas Experiment Station* 73, 48 pp.

Guhl, A.M. and Ortman, L.L. (1953) Visual patterns in the recognition of individuals among chickens. *Condor* 55, 287–297.

Guhl, A.M., Collias, N.E. and Allee, W.C. (1945) Mating behavior and the social hierarchy in small flocks of White Leghorns. *Physiological Zoology* 18, 365–390.

Gunnarsson, S., Keeling, L.J. and Svedberg, J. (1999) Effect of rearing factors on the prevalence of floor eggs, cloacal cannibalism and feather pecking in commercial flocks of loose housed hens. *British Poultry Science* 40, 12–18.

Gunnarsson, S., Yngvesson, J., Keeling, L.J. and Forkman, B. (2000) Rearing without access to perches impairs the spatial skills of laying hens. *Applied Animal Behaviour Science* 67, 217–228.

Guyomarc'h, C., Guyomarc'h, J.C. and Garnier, D.H. (1981) Influence of male vocalisations on the reproduction of quail females (*Coturnix coturnix japonica*). *Biology of Behaviour* 6, 167–182.

Guyomarc'h, J.C. (1967) Contribution à l'étude des cris de contact chez la caille japonaise (*Coturnix c. japonica*). *Comptes Rendus 91ème Congress Nationale des Sociétiés Savantes de Rennes* 3, 354–360.

Gvaryahu, G., Ararat, E., Asaf, E., Lev, M., Weller, J.L., Robinzon, B. and Snapir, N. (1994) An enrichment object that reduces aggressiveness and mortality in caged laying hens. *Physiology of Behavior* 55, 313–316.

Hale, E.B. (1955) Defects in sexual behavior as factors affecting fertility in turkeys. *Poultry Science* 34, 1059–1067.

Hale, E.B. (1962) Domestication and the evolution of behaviour. In: Hafez, E.S.E. (ed.) *The Behaviour of Domestic Animals*. Ballière Tindall, London, pp. 21–53.

Hale, E.B., Schleidt, W.M. and Schein, M.W. (1969) The behaviour of turkeys. In: Hafez, E.S.E. (ed.) *The Behaviour of Domestic Animals*, 2nd edn. Williams and Wilkins, Baltimore, Maryland, pp. 22–44.

Halpern, B.P. (1962) Gustatory nerve responses in the chicken. *American Journal of Physiology* 203, 541–544.

Hann, C.M. (1980) Some system definitions and characteristics. In: Moss, R. (ed.) *The Laying Hen and Its Environment*. Martinus Nijhoff, The Hague, The Netherlands, pp. 239–258.

Hansen, R.S. (1976) Nervousness and hysteria of mature female chickens. *Poultry Science* 55, 531–543.

Harrison, R. (1964) *Animal Machines*. Stuart, London.

Harrison, R. (1989) Research into action – some concerns. In: Faure, J.M. and Mills, A.D. (eds) *Proceedings, Third European Symposium on Poultry Welfare*. World's Poultry Science Association, Tours, France, pp. 253–255.

Harrison, R. (1991) The myth of the barn egg. *New Scientist* 30th November, 40–43.

Hart, B.L. (1988) Biological basis of the behavior of sick animals. *Neuroscience and Biobehavioral Reviews* 12, 123–137.

Health and Safety Executive (1980) *Threshold Limit Values*. HMSO, London.

Healy, W.M. and Thomas, J.W. (1973) Effects of dusting on the plumage of Japanese quail. *Wilson Bulletin* 442–448.

Hearn, P.J. (1976) A comparison of troughs, nipples and cup drinkers for laying hens in cages. *Gleadthorpe Experimental Husbandry Farm Poultry Booklet*. Gleadthorpe, UK, pp. 94–98.

Hearn, P.J. (1981) The effect of time of feeding and position of nest boxes on floor eggs. *MAFF/ADAS Report PH 03555*. Agricultural Development and Advisory Service, Nottingham, UK.

Hearn, P.J. and Gooderham, K.R. (1988) Ducks. In: Universities Federation for Animal Welfare (eds) *Management and Welfare of Farm Animals*. Ballière Tindall, London, pp. 243–253.

Heil, G., Simianer, H. and Dempfle, L. (1990) Genetic and phenotypic variation in prelaying behaviour of Leghorn hens kept in single cages. *Poultry Science* 69, 1231–1235.

Her Majesty's Stationery Office (1965) *Report of the Technical Committee to Enquire into the Welfare of Animals Kept Under Intensive Livestock Husbandry Systems*. Command Paper 2836. HMSO, London.

Her Majesty's Stationery Office (1987) *Animals, Prevention of Cruelty, The Welfare of Battery Hens Regulations 1987*. HMSO, London.

Hewson, P. (1986) Origin and development of the British poultry industry: the first hundred years. *British Poultry Science* 27, 525–539.

Hill, J.A. (1977) The relationship between food and water intake in the laying hen. PhD Thesis, Huddersfield Polytechnic, Huddersfield, UK.

Hill, J.A. (1983) Aviary system poses feather pecking and floor egg problems. *Poultry International* May, 109–113.

Hill, J.A. (1986) Egg production in alternative systems – a review of recent research in the UK. *Research and Development in Agriculture* 3, 13–18.

Hocking, P.M. and Duff, S.R.I. (1989) Musculo-skeletal lesions in adult male broiler breeder fowls and their relationships with body weight and fertility at 60 weeks of age. *British Poultry Science* 30, 777–784.

Hodgetts, B. (1981) *Dealing with Dirty Hatching Eggs.* Hatch Handout 17, MAFF Information for Flock Farms and Hatcheries.

Hogan, J.A. (1971) The development of a hunger system in young chicks. *Behaviour* 39, 128–201.

Hogan, J.A. (1973) Development of food recognition in young chicks. I. Maturation and nutrition. *Journal of Comparative and Physiological Psychology* 83, 355–356.

Hogan, J.A. and von Boxel, F. (1993) Causal factors controlling dustbathing in Burmese red junglefowl: some results and a model. *Animal Behaviour* 46, 627–635.

Holcombe, D.J., Roland, D.A. and Harms, R.H. (1976a) The ability of hens to regulate phosphorus intake when offered diets containing different levels of phosphorus. *Poultry Science* 55, 308–317.

Holcombe, D.J., Roland, D.A. and Harms, R.H. (1976b) The ability of hens to regulate protein intake when offered diets containing different levels of protein. *Poultry Science* 55, 1731–1737.

Huber, H., Fölsch, D.W. and Stahli, U. (1985) Influence of various nesting materials on nest-site selection of the domestic hen. *British Poultry Science* 26, 367–373.

Huber-Eicher, B. and Sebo, F. (2001a) Reducing feather pecking when raising laying hen chicks in aviary systems. *Applied Animal Behaviour Science* 73, 59–68.

Huber-Eicher, B. and Sebo, F. (2001b) The prevalence of feather pecking and development in commercial flocks of laying hens. *Applied Animal Behaviour Science* 74, 223–231.

Huber-Eicher, B. and Wechsler, B. (1998) The effect of quality and availability of foraging materials on feather pecking in laying chicks. *Animal Behaviour* 55, 861–873.

Hughes, B.O. (1971) Allelomimetic feeding in the domestic fowl. *British Poultry Science* 12, 359–366.

Hughes, B.O. (1972) A circadian rhythm of calcium intake in the domestic fowl. *British Poultry Science* 13, 485–493.

Hughes, B.O. (1975a) Spatial preference in the domestic hen. *British Veterinary Journal* 131, 560–564.

Hughes, B.O. (1975b) The concept of an optimal stocking density and its selection for egg production. In: Freeman, B.M. and Boorman, K.N. (eds) *Economic Factors Affecting Egg Production. Poultry Science Symposium 10.* British Poultry Science Ltd, Edinburgh, UK, pp. 271–298.

Hughes, B.O. (1976) Preference decisions of domestic hens for wire or litter floors. *Applied Animal Ethology* 2, 155–165.

Hughes, B.O. (1977) Selection of group size by individual laying hens. *British Poultry Science* 19, 9–18.

Hughes, B.O. (1983) Conventional and shallow cages: a summary of research from welfare and production aspects. *World's Poultry Science Journal* 39, 218–228.

Hughes, B.O. (1984) The principles underlying choice of feeding behaviour in fowls – with special reference to production experiments. *World's Poultry Science Journal* 40, 141–150.

Hughes, B.O. (1985) Feather loss as a problem: how does it occur? In: Wegner, R.M. (ed.) *Proceedings, Second European Seminar on Poultry Welfare.* World's Poultry Science Association, Celle, Germany, pp. 177–188.

Hughes, B.O. and Appleby, M.C. (1989) Increase in bone strength of spent laying hens housed in modified cages with perches. *Veterinary Record* 124, 483–484.

Hughes, B.O. and Black, A.J. (1973) The preference of domestic hens for different types of battery cage floor. *British Poultry Science* 14, 615–619.

Hughes, B.O. and Black, A.J. (1974) The effect of environmental factors on activity, selected behaviour patterns and 'fear' of fowls in cages and pens. *British Poultry Science* 15, 375–380.

Hughes, B.O. and Black, A.J. (1976) Battery cage shape: its effect on diurnal feeding pattern, egg shell cracking and feather pecking. *British Poultry Science* 17, 327–336.

Hughes, B.O. and Curtis, P.E. (1997) Health and disease. In: Appleby, M.C. and Hughes, B.O. (eds) *Animal Welfare*. CAB International, Wallingford, UK, pp. 109–125.

Hughes, B.O. and Dewar, W.A. (1971) A specific appetite for zinc in zinc-depleted domestic fowl. *British Poultry Science* 12, 225.

Hughes, B.O. and Dun, P. (1983) A comparison of laying stock housed intensively in cages and outside on range. Years 1981–1983. *Research and Development Publication No. 18*. West of Scotland Agricultural College, Ayr, UK.

Hughes, B.O. and Dun, P. (1986) A comparison of hens housed intensively in cages and outside on range. *Zootechnica International* February, 44–46.

Hughes, B.O. and Duncan, I.J.H. (1972) The influence of strain and environmental factors upon feather pecking and cannibalism in fowls. *British Poultry Science* 13, 525–547.

Hughes, B.O. and Gentle, M.J. (1995) Beak trimming of poultry: its implications for welfare. *World's Poultry Science Journal* 51, 51–61.

Hughes, B.O. and Michie, W. (1982) Plumage loss in medium bodied hybrid hens: the effect of beak trimming and cage design. *British Poultry Science* 23, 59–64.

Hughes, B.O. and Whitehead, C.C. (1979) Behavioural changes associated with the feeding of low-sodium diets to laying hens. *Applied Animal Ethology* 5, 255–266.

Hughes, B.O. and Wood-Gush, D.G.M. (1971) Investigations into specific appetites for sodium and thiamine in domestic fowls. *Physiology and Behavior* 6, 331–229.

Hughes, B.O. and Wood-Gush, D.G.M. (1977) Agonistic behaviour in domestic hens: the influence of housing method and group size. *Animal Behaviour* 25, 1056–1062.

Hughes, B.O., Gilbert, A.B. and Brown, M.F. (1986) Categorisation and causes of abnormal egg shells: relationship with stress. *British Poultry Science* 27, 325–337.

Hughes, B.O., Duncan, I.J.H. and Brown, M.F. (1989) The performance of nest building by domestic hens: is it more important than the construction of a nest? *Animal Behaviour* 37, 210–214.

Hughes, B.O., Carmichael, N.L., Walker, A.W. and Grigor, P.N. (1997) Low incidence of aggression in large flocks of laying hens. *Applied Animal Behaviour Science* 54, 215–234.

Huon, F., Meunier-Salaün, M.-C. and Faure, J.-M. (1986) Feeder design and available feeding space influence the feeding behaviour of hens. *Applied Animal Behaviour Science* 15, 65–70.

Hurnik, J.F. and Lehman, H. (1985) The philosophy of farm animal welfare: a contribution to the assessment of farm animal well-being. In: Wegner, R.M. (ed.) *Proceedings, Second European Symposium on Poultry Welfare*. World's Poultry Science Association, Celle, Germany, pp. 255–266.

Hurnik, J.F. and Lehman, H. (1988) Ethics and farm animal welfare. *Journal of Agricultural Ethics* 1, 305–318.

Inglis, I.R., Forkman, B. and Lazarus, J. (1997) Free food or earned food? A review and fuzzy model of contrafreeloading. *Animal Behaviour* 53, 1171–1191.

Jackson, S. and Duke, G.E. (1995) Intestine fullness influences feeding behaviour and crop filling in the domestic turkey. *Physiology and Behavior* 58, 1027–1034.

Jenner, T.D. and Appleby, M.C. (1991) Effects of space allowance on behavioural restriction and synchrony in hens. *Applied Animal Behaviour Science* 31, 292–293.

Jensen, P. and Toates, F.M. (1993) Who needs 'behavioural needs'? Motivational aspects of the needs of animals. *Applied Animal Behaviour Science* 37, 161–181.

Johnsen, P.F., Vestergaard, K.S. and Nørgaard-Nielsen, G. (1998) Influence of early rearing conditions on the development of feather pecking and cannibalism in domestic fowl. *Applied Animal Behaviour Science* 60, 25–41.

Johnsgard, P.A. (1973) *Grouse and Quails of North America*. University of Nebraska, Lincoln, Nebraska.

Johnson, S.B., Hamm, R.J. and Leahey, T.H. (1986) Observational learning in *Gallus gallus domesticus* with and without a conspecific model. *Bulletin of the Psychonomic Society* 24, 237–239.

Jones, E.K.M., Prescott, N.B., Cook, P., White, R.P. and Wathes, C.M. (2001) Ultraviolet light and mating behaviour in domestic broiler breeders. *British Poultry Science* 42, 23–32.

Jones, M.C. and Leighton, A.T. Jr (1987) Research note: effect of presence or absence of the opposite sex on egg production and semen quality of breeder turkeys. *Poultry Science* 66, 2056–2059.

Jones, M.E. and Mench, J.A. (1991) Behavioral correlates of male mating success in a multisire flock as determined by DNA fingerprinting. *Poultry Science* 70, 1493–1498.

Jones, R.B. (1982) Effects of early environmental enrichment upon open field behavior and timidity in the domestic chick. *Developmental Psychobiology* 15, 105–111.

Jones, R.B. (1986) The tonic immobility reaction of the domestic fowl: a review. *World's Poultry Science Journal* 42, 82–96.

Jones, R.B. (1987a) Fearfulness of caged laying hens: the effects of cage level and type of roofing. *Applied Animal Behaviour Science* 17, 171–175.

Jones, R.B. (1987b) Social and environmental aspects of fear in the domestic fowl. In: Zayan, R. and Duncan, I.J.H. (eds) *Cognitive Aspects of Social Behaviour in the Domestic Fowl*. Elsevier, Amsterdam, pp. 82–149.

Jones, R.B. (1989) Development and alleviation of fear. In: Faure, J.M. and Mills, A.D. (eds) *Proceedings, Third European Symposium on Poultry Welfare*. World's Poultry Science Association, Tours, France, pp. 123–136.

Jones, R.B. and Faure, J.M. (1981) Tonic immobility (righting time) in laying hens housed in cages and pens. *Applied Animal Ethology* 7, 369–372.

Jones, R.B. and Gentle, M.J. (1985) Olfaction and behavioural modification in domestic chicks (*Gallus domesticus*). *Physiology and Behavior* 34, 917–924.

Jones, R.B. and Hocking, P.M. (2000) Genetic selection for poultry behaviour: big bad wolf or friend in need? *Animal Welfare* 8, 343–359.

Jones, R.B., Duncan, I.J.H. and Hughes, B.O. (1981) The assessment of fear in domestic hens exposed to a looming human stimulus. *Behavioural Processes* 6, 121–133.

Jones, R.B., Mills, A.D. and Faure, J.M. (1991) Genetic and experiential manipulation of fear-related behavior in Japanese quail (*Coturnix c. japonica*). *Journal of Comparative Psychology* 105, 15–24.

Jordan, F.T.W. (ed.) (1990) *Poultry Diseases*, 3rd edn. Baillière Tindall, London.

Julian, R.J. (1995) Population dynamics and diseases of poultry. In: Hunton, P. (ed.) *Poultry Production*. Elsevier, Amsterdam, pp. 525–560.

Jull, M.A. (1938) *Poultry Husbandry*. McGraw Hill, New York.

Justice, W.P., McDaniel, G.R. and Craig, J.V. (1962) Techniques for measuring sexual effectiveness in male chickens. *Poultry Science* 41, 732–739.

Kant, I. (1786) *Grundlegung zur Metaphysik der Sitten*. Hartnoch, Riga, Latvia.

Keeling, L.J. (1994) Feather pecking – who in the group does it, how often and under what circumstances? In: *Proceedings of 9th European Poultry Conference*. Glasgow, UK, pp. 288–289.

Keeling, L.J. and Duncan, I.J.H. (1989) Interindividual distances and orientation in laying hens housed in groups of three in two different sized enclosures. *Applied Animal Behaviour Science* 24, 325–342.

Keeling, L.J. and Duncan, I.J.H. (1991) Social spacing in domestic fowl under semi-natural conditions. The effect of behavioural activity and activity transitions. *Applied Animal Behaviour Science* 32, 205–217.

Keeling, L.J., Hughes, B.O. and Dun, P. (1988) Performance of free range laying hens in a polythene house and their behaviour on range. *Farm Buildings Progress* 94, 21–28.

Kennedy, J.S. (1992) *The New Anthropomorphism*. Cambridge University Press, Cambridge.

Kestin, S.C. (1994) *Pain and Stress in Fish*. RSPCA, Horsham, UK.

Kestin, S.C., Knowles, T.G., Tinch, A.E. and Gregory, N.G. (1992) Prevalence of leg weakness in broiler chickens and its relationship with genotype. *Veterinary Record* 131, 190–194.

Kettlewell, P.J. and Mitchell, M.A. (2001) Comfortable ride: Concept 2000 provides climate control during poultry transport. *Resource Engineering and Technology for a Sustainable World* 8, 13–14.

Kiley-Worthington, M. (1989) Ecologicial, ethological and ethically sound environments for animals: towards symbiosis. *Journal of Agricultural Ethics* 2, 323–347.

King, A.S. and McLelland, J. (1975) *Outline of Avian Anatomy*. Baillière Tindall, London.

King, A.S. and McLelland, J. (1979–1985) *Form and Function in Birds*, Vols 1–3. Academic Press, London.

Kirk, S., Emmans, G.C., McDonald, R. and Arnist, D. (1980) Factors affecting the hatchability of eggs from broiler breeders. *British Poultry Science* 21, 37–43.

Kite, V.C., Cumming, R.B. and Wodzicka-Tomaszewska, M. (1980) Nesting behaviour of hens in relation to the problem of floor eggs. In: Wodzicka-Tomaszewska, M., Edey, T.N. and Lynch, J.J. (eds) *Behaviour in Relation to Reproduction, Management and Welfare of Farm Animals*. Reviews in Rural Science IV, Armidale, Australia, pp. 93–96.

Kjaer, J.B. and Mench, J.A. (2003) Behaviour problems associated with selection for increased production. In: Muir, W.M. and Aggrey, S.E. (eds) *Poultry Genetics, Breeding and Biotechnology*. CAB International, Wallingford, UK, pp. 67–82.

Klopfer, P.H. (1959) An analysis of learning in young Anatidae. *Ecology* 40, 90–102.

Knierim, U. and Jackson, W.T. (1997) Legislation. In: Appleby, M.C. and Hughes, B.O. (eds) *Animal Welfare*. CAB International, Wallingford, UK, pp. 249–264.

Knowles, T.G. and Broom, D.M. (1990) Limb bone strength and movement in laying hens from different housing systems. *Veterinary Record* 126, 354–356.

Knowles, T.G. and Wilkins, L.J. (1999) The problem of broken bones during the handling of laying hens – a review. *Poultry Science* 77, 1798–1802.

Koštál, L., Savory, C.J. and Hughes, B.O. (1992) Diurnal and individual variation in behaviour of restricted-fed broiler breeders. *Applied Animal Behaviour Science* 32, 361–374.

Kratzer, D.D. and Craig, J.V. (1980) Mating behavior of cockerels: effects of social status, group size and group density. *Applied Animal Ethology* 6, 49–62.

Kristensen, H.H., Burgess, L.R., Demmers, T.G.H. and Wathes, C.M. (2000) The preferences of laying hens for different concentrations of atmospheric ammonia. *Applied Animal Behaviour Science* 68, 307–318.

Kruijt, J.P. (1964) Ontogeny of social behaviour in the Burmese Red Junglefowl (*Gallus gallus spadecius* Bonaterre). *Behaviour* Supplement XII, 1–201.

Kyriazakis, I. and Savory, C.J. (1997) Hunger and thirst. In: Appleby, M.C. and Hughes, B.O. (eds) *Animal Welfare*. CAB International, Wallingford, UK, pp. 49–62.

Lacy, M.P. and Czarick, M. (1998) Mechanical harvesting of broilers. *Poultry Science* 77, 1794–1797.

Lake, P.E. (1969) Factors affecting fertility. In: Carter, T.C. and Freeman, B.M. (eds) *The Fertility and Hatchability of the Hen's Egg*. Oliver and Boyd, Edinburgh, UK, pp. 3–29.

Landauer, W. (1967) The hatchability of chicken eggs as influenced by environment and heredity. *Storrs Agricultural Experiment Station Monographs* (revised).

Latham, R.M. (1976) *Complete Book of the Wild Turkey*. Stackpole Books, Harrisburg, Pennsylvania.

Leahy, M.P.T. (1991) *Against Liberation: Putting Animals in Perspective*. Routledge, London.

Leonard, M.L. and Horn, A.G. (1995) Crowing in relation to status in roosters. *Animal Behaviour* 49, 1283–1290.

Leonard, M.L., Zanette, L. and Fairfull, R.W. (1993a) Early exposure to females affects interactions between male White Leghorn chickens. *Applied Animal Behaviour Science* 36, 29–38.

Leonard, M.L., Zanette, L., Thompson, B.K. and Fairfull, R.W. (1993b) Early exposure to the opposite sex affects mating behaviour in White Leghorn chickens. *Applied Animal Behaviour Science* 37, 57–67.

Lewis, N.J. and Hurnik, J.F. (1979) Stimulation of feeding in neonatal turkeys by flashing lights. *Applied Animal Ethology* 5, 161–171.

Lewis, P.D., Perry, G.C. and Tuddenham, A. (1987) Noise output of hens subjected to interrupted lighting regimens. *British Poultry Science* 28, 535–540.

Li, T., Howland, H.C. and Troilo, D. (2000) Diurnal illumination patterns affect the development of the chick eye. *Vision Research* 40, 2387–2393.

Lill, A. (1968) Spatial organization in small flocks of domestic fowl. *Behaviour* 22, 258–290.

Lindberg, A.C. and Nicol, C.J. (1996) Space and density effects on group size preferences in laying hens. *British Poultry Science* 37, 721–729.

Lindberg, A.C. and Nicol, C.J. (1997) Dustbathing in modified battery cages: is sham dustbathing an adequate substitute? *Applied Animal Behaviour Science* 55, 113–128.

Lintern-Moore, S. (1972) The relationship between water intake and the production of 'wet' droppings in the domestic fowl. *British Poultry Science* 13, 237–242.

Loeffler, K. (1986) Assessing pain by studying posture, structure and function. In: Duncan, I.J.H. and Molony, V. (eds) *Assessing Pain in Farm Animals*. Commission of the European Communities, Luxembourg, pp. 49–55.

Lucas, C. (2001) *Stopping the Great Food Swap*. The Greens, European Parliament, Brussels, Belgium.

Ludvigsen, J.B., Empel, J., Kovacs, F., Manfredini, M., Unshelm, J. and Viso, M. (1982) Animal health and welfare. *Livestock Production Science* 9, 65–87.

Lundberg, A. and Keeling, L.J. (1999) The impact of social factors on nesting in laying hens (*Gallus gallus domesticus*). *Applied Animal Behaviour Science* 64, 57–69.

Lynn, N.J. (1989) Effect of the degree and duration of different feeding regimens on laying hens undergoing an induced moult. *British Poultry Science* 30, 970.

Mace, R. (1995) Why do we do what we do? *Trends in Ecology and Evolution* 10, 4–5.

Maguire, S.N. (1986) A study of some factors influencing the nest-site selection of the domestic hen in relation to the problem of floor eggs. MSc Thesis, University of New England, Armidale, Australia.

Manser, C.E. (1996) Effects of lighting on the welfare of domestic poultry: a review. *Animal Welfare* 5, 341–360.

Market and Opinion Research International (1983) *Public Attitudes Towards Farmers*. Research Study Conducted for National Farmers' Union. Market and Opinion Research International, London.

Marler, P., Dufty, A. and Pickert, R. (1986) Vocal communication in the domestic chicken. II. Is a sender sensitive to the presence and nature of a receiver? *Animal Behaviour* 34, 194–198.

Marsden A., Morris, T.R. and Cromarty, A.S. (1987) Effects of constant environmental temperatures on the performance of laying pullets. *British Poultry Science* 28, 361–380.

Martin, J.D. (2000) Storm damages poultry houses in South. *Feedstuffs* 7th February, 3.

Martrenchar, A., Huonnic, D. and Cotte, J.P. (2001) Influence of environmental enrichment on injurious pecking and perching behaviour in young turkeys. *British Poultry Science* 42, 161–170.

Mashaly, M.M., Webb, M.L., Youtz, S.L., Roush, W.B. and Graves, H.B. (1984) Changes in serum corticosterone concentration of laying hens as a response to increased population density. *Poultry Science* 63, 2271–2274.

Matter, F. and Oester, H. (1989) Hygiene and welfare implications of alternative husbandry systems for laying hens. In: Faure, J.M. and Mills, A.D. (eds) *Proceedings, Third European Symposium on Poultry Welfare.* World's Poultry Science Association, Tours, France, pp. 201–212.

Matthews, L.R., Temple, W., Foster, T.M. and McAdie, T.M. (1993) Quantifying the environmental requirements of layer hens by behavioural demand functions. In: Nichelmann, M., Wierenga, H.K. and Braun, S. (eds) *Proceedings, International Congress on Applied Ethology.* Berlin, Germany, pp. 206–209.

Maxwell, M.H., Dick, L.A., Anderson, I.A. and Mitchell, M.A. (1989) Ectopic cartilaginous and osseous lung nodules induced in the young broiler by inadequate ventilation. *Avian Pathology* 18, 113–124.

May, G.C. and Hawksworth, D. (1982) *British Poultry Standards*, 4th edn. Butterworth Scientific, London.

McAdie, T.M. and Keeling, L.J. (2000) Effect of manipulating feathers of laying hens on the incidence of feather pecking and cannibalism. *Applied Animal Behaviour Science* 68, 215–229.

McBride, G. (1958) Relationship between aggressiveness and egg production in the domestic hen. *Nature* 181, 858.

McBride, G. (1970) The social control of behaviour in fowls. In: Freeman, B.M. and Gordon, R.F. (eds) *Aspects of Poultry Behaviour.* British Poultry Science, Edinburgh, UK, pp. 3–13.

McBride, G. and Foenander, F. (1962) Territorial behaviour in flocks of domestic fowls. *Nature* 194, 102.

McBride, G., Parer, I.P. and Foenander, F. (1969) The social organization and behaviour of the feral domestic fowl. *Animal Behaviour Monographs* 2, 125–181.

McDaniel, G.R. and Craig, J.V. (1959) Behavior traits, semen measurements and fertility of White Leghorn males. *Poultry Science* 38, 1005–1014.

McFerran, J.B. and Stuart, J.C. (1990) Adenoviruses. In: Jordan, F.T.W. (ed.) *Poultry Diseases*, 3rd edn. Baillière Tindall, London, pp. 182–193.

McGeown, D., Danbury, T.C., Waterman-Pearson, A.E. and Kestin, S.C. (1999) Effect of carprofen on lameness in chickens. *Veterinary Record* 144, 668–671.

McInerney, J.P. (1994) Animal welfare: an economic perspective. In: Bennett, R.M. (ed.) *Valuing Farm Animal Welfare.* Occasional Paper No. 3. Department of Agricultural Economics and Management, The University of Reading, Reading, UK, pp. 9–25.

McInerney, J.P. (1998) The economics of welfare. In: Michell, A.R. and Ewbank, R. (eds) *Ethics, Welfare, Law and Market Forces: the Veterinary Interface.* Universities Federation for Animal Welfare, Wheathampstead, UK, pp. 115–132.

McKeegan, D.E.F. and Savory, C.J. (1999) Feather eating in layer pullets and its possible role in the aetiology of feather pecking damage. *Applied Animal Behaviour Science* 65, 73–85.

McKinney, F. (1975) The behaviour of ducks. In: Hafez, E.S.E. (ed.) *The Behaviour of Domestic Animals*, 3rd edn. Ballière Tindall, London, pp. 490–519.

McLean, K.A., Baxter, M.R. and Michie, W. (1986) A comparison of the welfare of laying hens in battery cages and in a perchery. *Research and Development in Agriculture* 3, 93–98.

Meijsser, F.M. and Hughes, B.O. (1989) Comparative analysis of pre-laying behaviour in battery cages and in three alternative systems. *British Poultry Science* 30, 747–760.

Mench, J.A. (1988) The development of agonistic behavior in male broiler chicks: a comparison with laying-type males and the effects of feed restriction. *Applied Animal Behaviour Science* 21, 233–242.

Mench, J.A. (1995) Animal welfare and management issues associated with the use of artificial insemination for broiler breeders. In: Bakst, M.R. and Wishart, G.W. (eds) *Proceedings, First International Symposium on the Artificial Insemination of Poultry.* Poultry Science Association, Savoy, Illinois, pp. 159–175.

Mench, J.A. (1998) Thirty years after Brambell: whither animal welfare science? *Journal of Applied Animal Welfare Science* 1, 91–102.

Mench, J.A. (2002) Broiler breeders: feed restriction and welfare. *World's Poultry Science Journal* 58, 23–29.

Mench, J.A. (2003) Assessing animal welfare at the farm and group level: a US perspective. *Animal Welfare* 13, 493–503.

Mench, J.A. and Duncan, I.J.H. (1998) Poultry welfare in North America: opportunities and challenges. *Poultry Science* 77, 1763–1765.

Mench, J.A. and Keeling, L.J. (2001) The social behaviour of domestic birds. In: Keeling, L.J. and Gonyou, H.W. (eds) *Social Behaviour in Farm Animals.* CAB International, Wallingford, UK, pp. 177–210.

Mench, J.A., van Tienhoven, A., Marsh, J.A., McCormick, C.C., Cunningham, D.L. and Baker, R.C. (1986) Effects of cage and floor pen management on behavior, production and physiological stress responses of laying hens. *Poultry Science* 65, 1058–1069.

Mendl, M. and Newberry, R.C. (1997) Social conditions. In: Appleby, M.C. and Hughes, B.O. (eds) *Animal Welfare.* CAB International, Wallingford, UK, pp. 191–203.

Metheringham, J. (2000) Disposal of day-old chicks – the way forward. *World Poultry* 16, 25–27.

Michie, W. and Wilson, C.W. (1985) *The Perchery System for Housing Layers.* Scottish Agricultural Colleges Research and Development Note No. 25.

Midgley, M. (1986) Letter to the editors. *Between the Species* 2, 195–196.

Miller, D.B. (1978) Species-typical and individually distinctive acoustic features of the crow calls of red jungle fowl. *Zeitschrift für Tierpsychologie* 47, 182–193.

Miller, L. (1942) Some tagging experiments with black-footed albatrosses. *Condor* 44, 3–9.

Millman, S.T. (1999) An investigation into extreme aggressiveness of broiler breeder males. PhD Thesis, University of Guelph, Canada.

Millman, S.T. and Duncan, I.J.H. (2000a) Strain differences in aggressiveness of male domestic fowl in response to a male model. *Applied Animal Behaviour Science* 66, 217–233.

Millman, S.T. and Duncan, I.J.H. (2000b) Effect of male-to-male aggressiveness and feed-restriction during rearing on sexual behaviour and aggressiveness towards females by male domestic fowl. *Applied Animal Behaviour Science* 70, 63–82.

Millman, S.T., Duncan, I.J.H. and Widowski, T.M. (2000) Male broiler breeder fowl display high levels of aggression toward females. *Poultry Science* 79, 1233–1241.

Mills, A.D., Wood-Gush, D.G.M. and Hughes, B.O. (1985a) Genetic analysis of strain differences in pre-laying behaviour in battery cages. *British Poultry Science* 26, 182–197.

Mills, A.D., Duncan, I.J.H., Slee, G.S. and Clark, J.S.B. (1985b) Heart rate and laying behavior in two strains of domestic chicken. *Physiology and Behavior* 35, 145–147.

Mills, A.D., Beilharz, R.G. and Hocking, P.M. (1997a) Genetic selection. In: Appleby, M.C. and Hughes, B.O. (eds) *Animal Welfare.* CAB International, Wallingford, UK, pp. 219–231.

Mills, A.D., Crawford, L.L., Domjan, M. and Faure, J.M. (1997b) The behavior of the Japanese or domestic quail *Coturnix japonica*. *Neuroscience and Biobehavioral Reviews* 21, 261–281.

Ministry of Agriculture, Fisheries and Food (1969) *Codes of Recommendations for the Welfare of Livestock: Domestic Fowls*. MAFF Publications Office, London.

Ministry of Agriculture, Fisheries and Food (1987a) *Codes of Recommendations for the Welfare of Livestock: Domestic Fowls*. MAFF Publications Office, London.

Ministry of Agriculture, Fisheries and Food (1987b) *Codes of Recommendations for the Welfare of Livestock: Turkeys*. MAFF Publications Office, London.

Ministry of Agriculture, Fisheries and Food (1997) *News Release*. 18th June, London.

Mitchell, M.A. and Kettlewell, P.J. (1993) Catching and transport of broiler chickens. In: Savory, C.J. and Hughes, B.O. (eds) *Proceedings, Fourth European Symposium on Poultry Welfare*. Universities Federation for Animal Welfare, Potters Bar, UK, pp. 219–229.

Mitchell, M.A. and Kettlewell, P.J. (1998) Physiological stress and welfare of broiler chickens in transit – solutions, not problems! *Poultry Science* 77, 1803–1814.

Moberg, G.P. (1999) When does stress become distress? *Lab Animal* 28, 22–26.

Moberg, G.P. (2000) Biological response to stress: implications for animal welfare. In: Moberg, G.P. and Mench, J.A. (eds) *The Biology of Animal Stress: Basic Principles and Implications for Animal Welfare*. CAB International, Wallingford, UK, pp. 1–21.

Mongin, P. and Saveur, B. (1974) Voluntary food and calcium intake by the laying hen. *British Poultry Science* 15, 349–359.

Mongin, P. and Saveur, B. (1979) The specific calcium appetite of the domestic fowl. In: Borman, K.M. and Freeman, B.M. (eds) *Food Intake Regulation in Poultry*. British Poultry Science Ltd, Edinburgh, UK, pp. 171–189.

Morishita, T.Y., Aye, P.P. and Harr, B.S. (1999) Crop impaction resulting from feather ball formation in caged layers. *Avian Diseases* 43, 160–163.

Morris, T.R. (1967) Light requirements of the fowl. In: Carter, C.T. (ed.) *Environmental Control in Poultry Production*. Oliver and Boyd, Edinburgh, UK, pp. 15–39.

Muir, W.M. (1996) Group selection for adaptation to multi-hen cages: selection program and direct responses. *Poultry Science* 75, 447–458.

Muir, W.M. and Craig, J.V. (1998) Improving animal well-being through genetic selection. *Poultry Science* 77, 1781–1788.

National Animal Health Monitoring System (2000) *Part II: Reference of 1999 Table Egg Layer Management in the U.S.* USDA, Washington, DC.

National Farmers' Union (1990) *Poultry, Some Facts and Figures*. National Farmers Union, London.

National Opinion Polls (1983) *Animal Issues and their Influence on Voting*. National Opinion Polls Market Research Ltd, London.

National Research Council (2002) *Frontiers in Agricultural Research: Food, Health, Environment and Communities*. National Academies Press, Washington, DC.

Necker, R. (2000) The somatosensory system, In: Whittow, G.C. (ed.) *Sturkie's Avian Physiology*, 5th edn. Academic Press, San Diego, California, pp. 57–69.

Nestle, M. (2002) *Food Politics: How the Food Industry Influences Nutrition and Health*. University of California Press, Berkeley, California.

Neuhaus, W. (1963) On the olfactory sense of birds. In: Zotterman, Y. (ed.) *Olfaction and Taste*. Pergamon, Oxford, pp. 111–123.

Newberry, R.C. (1995) Environmental enrichment – increasing the biological relevance of captive environments. *Applied Animal Behaviour Science* 44, 229–243.

Newberry, R.C. (1999) Exploratory behaviour of young domestic fowl. *Applied Animal Behaviour Science* 63, 311–321.

Newberry, R.C. and Shackleton, D.M. (1997) Use of visual cover by domestic fowl: a Venetian blind effect? *Animal Behaviour* 54, 387–395.

Newberry, R.C., Webster, B.A., Lewis, N.J. and Van Arnam, C. (1999) Management of spent hens. *Journal of Applied Animal Welfare Science* 2, 13–30.

Nicol, C.J. (1987a) Effect of cage height and area on the behaviour of hens housed in battery cages. *British Poultry Science* 28, 327–335.

Nicol, C.J. (1987b) Behavioural responses of laying hens following a period of spatial restriction. *Animal Behaviour* 35, 1709–1719.

Nicol, C.J. and Weeks, C. (2000) Poultry handling and transport. In: Grandin, T. (ed.) *Livestock Handling and Transport*, 2nd edn. CAB International, Wallingford, UK, pp. 363–384.

Nicol, C.J., Gregory, N.G., Knowles, T.G., Parkman, I.D. and Wilkins, L.J. (1999) Differential effects of increased stocking density, mediated by increased flock size, on feather pecking and aggression in laying hens. *Applied Animal Behaviour Science* 65, 137–152.

Nørgaard-Nielsen, G. (1990) Bone strength of laying hens kept in an alternative system, compared with hens in cages and on deep litter. *British Poultry Science* 31, 81–89.

Nørgaard-Nielsen, G. (1997) Dustbathing and feather pecking in domestic chickens reared with and without access to sand. *Applied Animal Behaviour Science* 52, 99–108.

Odén, K., Vestergaard, K.S. and Algers, B. (1999) Agonistic behavior and feather pecking in single-sexed and mixed groups of laying hens. *Applied Animal Behaviour Science* 62, 219–231.

Odén, K., Vestergaard, K.S. and Algers, B. (2000) Space use and agonistic behaviour in relation to sex composition in large flocks of laying hens. *Applied Animal Behaviour Science* 67, 307–320.

Odén, K., Keeling, L.J. and Algers, B. (2002) Behaviour of laying hens in two types of aviary systems on 25 commercial farms in Sweden. *British Poultry Science* 43, 169–181.

Odum, T.W., Wideman, R.F. and Coello, C.L. (1987) Current research on body fluid accumulation in broilers (ascites). *Zootechnica International* August, 53–54.

Oester, H. (1986) Systemes de détention récents pour pondeuses en Suisse. In: Larbier, M. (ed.) *Proceedings, Seventh European Poultry Conference*. World's Poultry Science Association, Paris, pp. 1077–1081.

Oester, H. and Fröhlich, E. (1989) Alternative systems in Switzerland. In: Commission of the European Communities (ed.) *Alternative Improved Housing Systems for Poultry*. Commission of the European Communities, Beekbergen, The Netherlands, pp. 50–58.

Olsson, I.A.S. and Keeling, L.J. (2002) The push-door for measuring motivation in hens: laying hens are motivated to perch at night. *Animal Welfare* 11, 11–19.

Ostfeld, R.S. and Lewis, D.N. (1999) Experimental studies of interactions between wild turkeys and black-legged ticks. *Journal of Vector Ecology* 24, 182–186.

Ottinger, M.A. and Mench, J.A. (1989) Reproductive behaviour in poultry: implications for artificial insemination technology. *British Poultry Science* 30, 431–442.

Ottinger, M.A., Schleidt, W.M. and Russek, E. (1982) Daily patterns in courtship and mating behavior in the male Japanese quail. *Behavioural Processes* 7, 223–233.

Ouart, M.D., Russell, G.B. and Wilson, H.R. (1989) Mating behavior in response to toe nail removal in broiler breeders. *Zootechnica International* 7, 35–37.

Oyetunde, O.O.F., Thomson, R.G. and Carlson, H.C. (1978) Aerosol exposure of ammonia, dust and *Escherichia coli* in broiler chickens. *Canadian Veterinary Journal* 19, 167–193.

Pagel, M. and Dawkins, M.S. (1997) Peck orders and group size in laying hens: 'futures contracts' for non-aggression. *Behavioral Processes* 40, 13–25.

Pamment, P., Foenander, F. and McBride, G. (1983) Social and spatial organization of male behaviour in mated domestic fowl. *Applied Animal Ethology* 9, 341–349.

Pattison, M. (ed.) (1993) *The Health of Poultry*. Longman, Harlow, UK.

Pepperberg, I. (1987) Evidence for conceptual quantitative abilities in the African grey parrot: labeling of cardinal sets. *Ethology* 75, 37–61.

Perry, G.C., Charles, D.R., Day, P.J., Hartland, J.R. and Spencer, P.G. (1971) Egg laying behaviour in a broiler parent flock. *World's Poultry Science Journal* 27, 162.

Perry, G.C., Stevens, K. and Allen, J. (1976) Particle selection by caged layers and pullets. In: *Proceedings, Fifth European Poultry Conference*. World's Poultry Science Association, Malta, pp. 1089–1096.

Petherick, J.C. and Rutter, S.M. (1990) Quantifying motivation using a computer-controlled push-door. *Applied Animal Behaviour Science* 27, 159–167.

Petherick, J.C., Waddington, D. and Duncan, I.J.H. (1990) Learning to gain access to a foraging and dustbathing substrate by domestic fowl: is 'out of sight out of mind'? *Behavioural Processes* 22, 213–226.

Petit-Riley, R. and Estevez, I. (2001) Effects of density on perching behavior of broiler chickens. *Applied Animal Behaviour Science* 71, 127–140.

Pizzari, T. and Birkhead, T.R. (2000) Female feral fowl eject sperm of subdominant males. *Nature* 405, 787–789.

Policy Commission on the Future of Farming and Food (2002) *Farming and Food: a Sustainable Future*. Cabinet Office, London.

Potash, L.M. (1970) Vocalizations elicited by electrical brain stimulation in *Coturnix coturnix japonica*. *Behaviour* 36, 149–167.

Prescott, N.B. and Wathes, C.M. (2002) Preference and motivation of laying hens to eat under different illuminances and the effect of illuminance on eating behaviour. *British Poultry Science* 43, 190–195.

Preston, A.P. and Mulder, J. (1989) Effect of vertical food-trough dividers on the feeding and agonistic behaviours of laying hens. *British Poultry Science* 30, 519–532.

Preston, A.P. and Murphy, L.B. (1989) Movement of broiler chickens reared in commercial conditions. *British Poultry Science* 30, 519–532.

Price, E.O. (2002) *Animal Domestication and Behavior*. CAB International, Wallingford, UK.

Proudfoot, F.G. and Hulan, H.W. (1985) Effects of stocking density on the incidence of scabby hip syndrome among broiler chickens. *Poultry Science* 64, 2001–2003.

Raj, M. (1998) Welfare during stunning and slaughter of poultry. *Poultry Science* 77, 1815–1819.

Rappaport, S. and Soller, M. (1966) Mating behavior, fertility and rate-of-gain in Cornish males. *Poultry Science* 45, 997–1003.

Regan, T. (1983) *The Case for Animal Rights*. University of California Press, Berkeley, California.

Reilly, W.M., Koelkebeck, K.W. and Harrison, P.C. (1991) Performance evaluation of heat-stressed commercial broilers provided water-cooled perches. *Poultry Science* 70, 1699–1703.

Reiss, M.J. and Straughan, R. (1996) *Improving Nature? The Science and Ethics of Genetic Engineering*. Cambridge University Press, Cambridge.

Rietveld-Piepers, B., Blokhuis, H.J. and Wiepkema, P.R. (1985) Egg-laying behaviour and nest-site selection of domestic hens kept in small floor pens. *Applied Animal Behaviour Science* 14, 75–88.

Roberts, D. and Farrar, J. (1993) *The Economics of Egg Production 1992*. Special Studies in Agricultural Economics, Report no. 22. Department of Agricultural Economics, University of Manchester, Manchester, UK.

Robertson, E.S., Appleby, M.C., Hogarth, G.S. and Hughes, B.O. (1989) Modified cages for laying hens: a pilot trial. *Research and Development in Agriculture* 6, 107–114.

Rogers, C.S., Appleby, M.C., Keeling, L., Robertson, E.S. and Hughes, B.O. (1989) Assessing public opinion on commercial methods of egg production: a pilot study. *Research and Development in Agriculture* 6, 19–24.

Rogers, L.J. (1995) *The Development of Brain and Behaviour in the Chicken*. CAB International, Wallingford, UK.

Rollin, B.E. (1993) Animal production and the new social ethic for animals. In: Baumgardt, B. and Gray, H.G. (eds) *Food Animal Well-being*. Purdue University, West Lafayette, Indiana, pp. 3–13.

Rollin, B.E. (1995) *The Frankenstein Syndrome: Ethical and Social Issues in the Genetic Engineering of Animals*. Cambridge University Press, Cambridge.

Roush, W.B. (1986) A decision analysis approach to the determination of population density in laying cages. *World's Poultry Science Journal* 42, 26–31.

Roush, W.B., Mashaly, M.M. and Graves, H.B. (1984) Effect of increased bird population in a fixed cage area on production and economic responses of single comb White Leghorn laying hens. *Poultry Science* 63, 45–48.

Rowland, K.W. (1985) Intermittent lighting for laying fowls: a review. *World's Poultry Science Journal* 41, 5–19.

Rowland, L.O., Fry, J.L., Christmas, R.B., O'Sheen, A.W. and Harris, R.H. (1972) Differences in tibia strength and bone ash among strains of layers. *Poultry Science* 51, 1612–1615.

Royal Society for the Prevention of Cruelty to Animals (1996) *Welfare Standards for Laying Hens*. RSPCA, Horsham, UK.

Royal Society for the Prevention of Cruelty to Animals (1997) *Welfare Standards for Laying Hens*. RSPCA, Horsham, UK.

Rozin, P. (1976) The selection of foods by rats, humans and animals. In: Rosenblatt, J.S., Hinde, R.A., Shaw, E. and Beer, C. (eds) *Advances in the Study of Behaviour 6*. Academic Press, New York, p. 21.

Rushen, J. (1982) The peck orders of chickens: how do they develop and why are they linear? *Animal Behaviour* 30, 1129–1137.

Ruszler, P.L. (1998) Health and husbandry considerations of induced moulting. *Poultry Science* 77, 1789–1793.

Rutter, S.M. and Duncan, I.J.H. (1989) Behavioural measures of aversion in domestic fowl. In: Faure, J.M. and Mills, A.D. (eds) *Proceedings, Third European Symposium on Poultry Welfare*. World's Poultry Science Association, Tours, France, pp. 277–279.

Rutter, S.M. and Duncan, I.J.H. (1991) Shuttle and one-way avoidance as measures of aversion in the domestic fowl. *Applied Animal Behaviour Science* 30, 117–124.

Ryder, R.D. (2000) *Animal Revolution: Changing Attitudes Towards Speciesism*. Berg, Oxford.

Sainsbury, D.W.B. (1971) Domestic fowls. In: *The UFAW Handbook on the Care and Management of Farm Animals*. Churchill and Livingstone, Edinburgh, UK, pp. 99–130.

Sainsbury, D.W.B. (1980) *Poultry Health and Management*. Granada, London.

Sainsbury, D.W.B. (2000) *Poultry Health and Management: Chickens, Ducks, Turkeys, Geese, Quail*, 4th edn. Blackwell, Oxford.

Sambraus, H.H. (1994) Das sexualverhalten des Afrikanishcen Stausses (*Struthio camelus*). *Tierärztliche Umschau* 50, 108–111.

Sandercock, D.A., Hunter, R.R., Mitchell, M.A. and Hocking, P.M. (2001) The effect of genetic selection for muscle yield on idiopathic myopathy in poultry: implications for welfare. In: Oester, H. and Wyss, C. (eds) *Proceedings, Sixth European Symposium on Poultry Welfare*. World's Poultry Science Association, Zollikofen, Switzerland, pp. 118–123.

Sandøe, P., Crisp, R. and Holtug, N. (1997) Ethics. In: Appleby, M.C. and Hughes, B.O. (eds) *Animal Welfare*. CAB International, Wallingford, UK, pp. 3–17.

Sanotra, G.S. (1999) Registrering af aktuel benstyrke hos slagtekyllinger (Velfaerdsmoniteringsprojekt). Dyrenes Beskyttelse, Frederiksberg, Denmark.

Sanotra, G.S., Damkjer-Lund, J. and Vestergaard, K.S. (2002) Influence of light–dark schedules and stocking density on behaviour, risk of leg problems and occurrence of chronic fear in broilers. *British Poultry Science* 43, 344–354.

Sansolini, A. (1999) Europe in brief. *Agscene* 134, 13.

Savenjie, B., Lambooij, E., Gerritzen, M.A. and Korf, J. (2002) Development of brain damage as measured by brain impedance recording, and changes in heart rate, and blood pressure induced by different stunning and killing methods. *Poultry Science* 81, 572–578.

Savory, C.J. (1976) Effects of different lighting regimes on diurnal feeding patterns of the domestic fowl. *British Poultry Science* 17, 341–350.

Savory, C.J. (1978) The relationship between food and water intake and the effects of water restriction on laying Brown Leghorn hens. *British Poultry Science* 19, 631–641.

Savory, C.J. (1979) Feeding behaviour. In: Borman, K.M. and Freeman, B.M. (eds) *Food Intake Regulation in Poultry.* British Poultry Science Ltd, Edinburgh, UK, pp. 277–323.

Savory, C.J. (1980) Meal occurrence in Japanese quail in relation to particle size and nutrient density. *Animal Behaviour* 28, 160–171.

Savory, C.J. (1982) Effects of broiler companions on early performance of turkeys. *British Poultry Science* 23, 81–88.

Savory, C.J. (1985) An investigation into the role of the crop in control of feeding in Japanese quail and domestic fowls. *Physiology and Behaviour* 35, 917–928.

Savory, C.J. (1989) The importance of invertebrate food to chicks of gallinaceous species. *Proceedings of the Nutrition Society* 48, 113–133.

Savory, C.J. and Lariviere, J.-M. (2000) Effects of qualitative and quantitative food restriction treatments on feeding motivational state and general activity level of growing broiler breeders. *Applied Animal Behaviour Science* 69, 135–147.

Savory, C.J. and Mann, J.S. (1997) Development of pecking damage in growing bantams in relation to floor litter substrate and plumage colour. *British Poultry Science* Supplement, S13–S14.

Savory, C.J., Wood-Gush, D.G.M. and Duncan, I.J.H. (1978) Feeding behaviour in a population of domestic fowls in the wild. *Applied Animal Ethology* 4, 13–27.

Savory, C.J., Seawright, E. and Watson, A. (1992) Stereotyped behaviour in broiler breeders in relation to husbandry and opioid receptor blockade. *Applied Animal Behaviour Science* 32, 349–360.

Savory, C.J., Maros, K. and Rutter, S.M. (1993) Assessment of hunger in growing broiler breeders in relation to a commercial restricted feeding programme. *Animal Welfare* 2, 131–152.

Savory, C.J., Hocking, P.M., Mann, J.S. and Maxwell, M. (1996) Is broiler breeder welfare improved by using qualitative rather than quantitative food restriction to limit growth rate? *Animal Welfare* 5, 105–127.

Scheele, C.W. (1997) Pathological changes in metabolism of poultry related to increasing production levels. *Veterinary Quarterly* 19, 127–130.

Schlosser, E. (2001) *Fast Food Nation: the Dark Side of the All-American Meal.* Houghton Mifflin, Boston, Massachusetts.

Schmidt-Nielsen, K. (1975) Recent advances in avian respiration. In: Peaker, M. (ed.) *Symposium of the Royal Society of London, No. 35, Avian Physiology.* Academic Press, London, pp. 33–47.

Schorger, A.W. (1966) *The Wild Turkey. Its History and Domestication.* University of Oklahoma Press, Norman, Oklahoma.

Schutz, F. (1965) Homosexualität und Prägung: eine experimentelle Untersuchung an Enten. *Psychologische Forschung* 28, 439–463.

Schutz, K.E. and Jensen, P. (2001) Effects of resource allocation on behavioural strategies: a comparison of red junglefowl (*Gallus gallus*) and two domesticated breeds of poultry. *Ethology* 107, 753–765.

Schutz, K.E., Forkman, B. and Jensen, P. (2001) Domestication effects on foraging strategy, social behaviour and different fear responses: a comparison between red junglefowl (*Gallus gallus*) and a modern layer strain. *Applied Animal Behaviour Science* 74, 1–14.

Scientific Committee on Animal Health and Welfare (2000) *The Welfare of Chickens Kept for Meat Production (Broilers)*. European Commission, Brussels, Belgium.

Scientific Veterinary Committee (1996) *Report on the Welfare of Laying Hens*. Commission of the European Communities Directorate-General for Agriculture VI/B/II.2, Brussels, Belgium.

Scott, G.B. and Parker, C.L. (1994) The ability of hens to negotiate between horizontal perches. *Applied Animal Behaviour Science* 42, 121–127.

Selye, H. (1932) The general adaptation syndrome and the diseases of adaptation. *Journal of Clinical Endocrinology* 6, 117–152.

Sen, A. (1987) *On Ethics and Economics*. Blackwell, Oxford.

Sewell, R.B.S. and Guha, B.S. (1931) Zoological remains. In: Marshall, J. (ed.) *Mohenjo-Daro and the Indus Civilisation II*. Arthur Probsthain, London, pp. 649–673.

Shanaway, M.M. (1994) *Quail Production Systems. A Review.* Food and Agriculture Organization of the United Nations, Rome, Italy.

Shanaway, M.M. (1999) Review. In: *Ostrich Production Systems. FAO Paper 144.* Food and Agriculture Organization, Rome, Italy.

Shea-Moore, M.M., Mench, J.A. and Thomas, O.P. (1990) The effect of dietary tryptophan on aggressive behavior in developing and mature broiler breeder males. *Poultry Science* 69, 1664–1669.

Sherwin, C.M. (ed.) (1994) *Modified Cages for Laying Hens.* Universities Federation for Animal Welfare, Potters Bar, UK.

Sherwin, C.M. and Devereux, C.L. (1999) Preliminary investigations of ultraviolet-induced markings on domestic turkey chicks and a possible role in injurious pecking. *British Poultry Science* 40, 429–433.

Sherwin, C.M. and Kelland, A. (1998) Time budgets, comfort behaviours and injurious pecking of turkeys housed in pairs. *British Poultry Science* 39, 325–332.

Sherwin, C.M. and Nicol, C.J. (1993) Factors affecting floor laying by hens in modified cages. *Applied Animal Behaviour Science* 36, 211–222.

Shields, S.J., Garner, J.P. and Mench, J.A. (2004) Dustbathing by broiler chickens: a comparison of preference for four different substrates. *Applied Animal Behaviour Science* 87, 69–82.

Siegel, H.S. and Siegel, P.B. (1961) The relationship of social competition with endocrine weights and activity in male chickens. *Animal Behaviour* 9, 151–158.

Siegel, P.B. (1965) Genetics of behavior: selection for mating ability in chickens. *Genetics* 52, 1269–1277.

Siegel, P.B. and Beane, W.L. (1963) Semen characteristics of chickens maintained in all-male flocks and in individual cages. *Poultry Science* 42, 1028–1030.

Siegel, P.B. and Siegel, H.S. (1964) Rearing methods and subsequent sexual behavior of male chickens. *Animal Behaviour* 1, 270–271.

Siegel, P.B., Phillips, R.E. and Folsom, E.F. (1965) Genetic variation in the crow of adult chickens. *Behaviour* 24, 229–235.

Siller, W.G. (1985) Deep pectoral myopathy: a penalty of successful selection for muscle growth. *Poultry Science* 64, 1591–1595.

Singer, P. (1975) *Animal Liberation.* New York Review of Books, New York.

Skewes, P.A. and Wilson, H.R. (1990) *Bobwhite Quail Production.* Pamphlet.

Smith, F.M., West, N.H. and Jones, D.R. (2000) The cardiovascular system. In: Whittow, G.C. (ed.) *Sturkie's Avian Physiology*, 5th edn. Academic Press, San Diego, California, pp. 141–123.

Smith, P. and Daniel, C. (1982) *The Chicken Book*. North Point Press, San Francisco, California.

Smith, W.K. (1981) Poultry housing problems in the tropics and subtropics. In: Clark, J.A. (ed.) *Environmental Aspects of Housing for Animal Production*. Butterworths, London, pp. 235–258.

Smith, W.K. and Dun, P. (1983) 'What type of nest?' Unpublished paper presented to British Poultry Breeders and Hatcheries Association Flock Farmer's Conference.

Social Surveys (1968) *Factory Farming – What the Farmers Really Think*. Survey carried out for the Coordinating Committee on Factory Farming. Social Surveys (Gallup Poll) Ltd, Newport, Isle of Wight, UK.

Soley, J.T. and Groenewald, H.B. (1999) Reproduction. In: Deeming, D.C. (ed.) *The Ostrich: Biology, Production and Health*. CAB International, Wallingford, UK, pp. 129–158.

Solomon, S.E. (1991) *Egg and Eggshell Quality*. Wolfe Publishing, London.

Sorensen, P. (1989) Broiler selection and welfare. In: Faure, J.M. and Mills, A.D. (eds) *Proceedings, Third European Symposium on Poultry Welfare*. World's Poultry Science Association, Tours, France, pp. 45–58.

Sossinka, R. (1982) Domestication in birds. In: Farner, D.S., King, J.R. and Parkes, K.C. (eds) *Avian Biology*, Vol. 6. Academic Press, New York, pp. 373–403.

Suboski, M.D. (1987) Environmental variables and releasing-valence transfer in stimulus-directed pecking of chicks. *Behavioural and Neural Biology* 47, 262–274.

Swanson, J.C. and Mench, J.A. (2000) Animal welfare: consumer viewpoints. *Proceedings U.C. Poultry Symposium and Egg Processing Workshops*, Modesto and Riverside, California. Available at: http://animalscience.ucdavis.edu/Avian/Swanson.pdf

Sykes, A. (1988) Intake of sodium following sodium deficiency in the laying hen. *British Poultry Science* 29, 884–885.

Tablante, N.L., Vaillancourt, J.P., Martin, S.W., Shoukri, M. and Estevez, I. (2000) Spatial distribution of cannibalism mortalities in commercial laying hens. *Poultry Science* 79, 705–708.

Tanaka, T. and Hurnik, J.F. (1992) Comparison of behavior and performance of laying hens housed in battery cages and an aviary. *Poultry Science* 71, 235–243.

Tannenbaum, J. (1995) *Veterinary Ethics*. Mosby-Year Book, Inc., St Louis, Missouri.

Tauson, R. (1984) Effects of a perch in conventional cages for laying hens. *Acta Agriculturae Scandinavica* 34, 193–209.

Tauson, R. (1985) Mortality in laying hens caused by differences in cage design. *Acta Agriculturae Scandinavica* 35, 165–174.

Tauson, R. (1986) Avoiding excessive growth of claws in caged laying hens. *Acta Agriculturae Scandinavica* 36, 95–106.

Tauson, R. (1988) Effects of redesign. In: Cambridge Poultry Conference (ed.) *Cages for the Future*. Agricultural Development and Advisory Service, Nottingham, UK, pp. 42–69.

Tauson, R. (1989) Cages for laying hens: yesterday and today . . . tomorrow? In: Faure, J.M. and Mills, A.D. (eds) *Proceedings, Third European Symposium on Poultry Welfare*. World's Poultry Science Association, Tours, France, pp. 165–181.

Tauson, R. (2000) Furnished cages and aviaries: production and health. In: *Proceedings, 21st World Poultry Congress*, Montreal, pp. 8–17.

Tauson, R. and Holm, K.E. (2001) First furnished small group cages for laying hens in evaluation program on commercial farms in Sweden. In: Oester, H. and Wyss, C. (eds) *Proceedings, Sixth European Symposium on Poultry Welfare*. World's Poultry Science Association, Zollikofen, Switzerland, pp. 26–32.

Temple, W., Foster, T.M. and O'Donnell, C.S. (1984) Behavioural estimates of auditory thresholds in hens. *British Poultry Science* 25, 487–493.

Te Velde, H., Aarts, N. and van Woerken, C. (2002) Dealing with ambivalence: farmers' and consumers' perceptions of animal welfare in livestock breeding. *Journal of Agricultural and Environmental Ethics* 15, 203–219.

Thomson, A.L. (1964) *A New Dictionary of Birds*. Nelson, London.

Thorpe, W.H. (1951) The definition of terms used in animal behaviour studies. *Bulletin of Animal Behaviour* 9, 34–40.

Tinbergen, N. (1963) On aims and methods of ethology. *Zeitschrift für Tierpschologie* 20, 410–433.

Todd, P. (1989) The Protection of Animals Acts 1911–1964. In: Blackman, D.E., Humphries, P.N. and Todd, P. (eds) *Animal Welfare and the Law*. Cambridge University Press, Cambridge, pp. 13–36.

Tolman, C.W. and Wilson, G.F. (1965) Social feeding in domestic chicks. *Animal Behaviour* 13, 134–142.

Tucker, S. (1989) Alternatives? *ADAS Poultry Journal* 3, 15–30.

United Egg Producers (2002) *Animal Husbandry Guidelines for U.S. Egg Laying Flocks*. UEP, Alpharetta, Georgia.

Upp, W. (1928) Preferential mating of fowls. *Poultry Science* 35, 969–976.

van Enckevort, J.W.F. (1965) Het werkgebied bij leghennen. *Veeteelt en Zuivelberichten* 530–536.

van Kampen, M. (1981) Thermal influences on poultry. In: Clark, J.A. (ed.) *Environmental Aspects of Housing for Animal Production*. Butterworths, London, pp. 131–147.

van Liere, D.W. (1991) Function and organization of dustbathing in laying hens. PhD Thesis, University of Wageningen, Wageningen, The Netherlands.

van Liere, D.W., Kooijman, J. and Wiepkema, P.R. (1990) Dustbathing behaviour of laying hens as related to quality of dustbathing material. *Applied Animal Behaviour Science* 26, 127–141.

Verhoog, H. and Visser, T. (1997) A view of intrinsic value not based on animal consciousness. In: Dol, M., Kasanmoentalib, S., Lijmbach, S., Rivas, E. and van den Bos, R. (eds) *Animal Consciousness and Animal Ethics*. Van Gorcum, Assen, The Netherlands, pp. 223–232.

Verwoerd, D.J., Deeming, D.C., Angel, C.R. and Perelman, B. (1999) Rearing environments around the world. In: Deeming, D.C. (ed.) *The Ostrich: Biology, Production and Health*. CAB International, Wallingford, UK, pp. 191–216.

Vestergaard, K.S., Hogan, J.A. and Kruijt, J.P. (1990) The development of a behaviour system: dustbathing in the Burmese red junglefowl: I. The influence of the rearing environment on the organization of dustbathing. *Behaviour* 112, 99–116.

Vestergaard, K.S., Kruijt, J.P. and Hogan, J.A. (1993) Feather pecking and chronic fear in groups of red junglefowl: their relations to dustbathing, rearing environment and social status. *Animal Behaviour* 45, 1127–1140.

Vestergaard, K.S., Skadhauge, E. and Lawson, L.G. (1997) The stress of not being able to perform dustbathing in laying hens. *Physiology and Behavior* 62, 413–419.

Vestergaard, K.S., Damm, B.I., Abbott, U.K. and Blidsoe, M. (1999) Regulation of dustbathing in feathered and featherless domestic chicks: the Lorenzian model revisited. *Animal Behaviour* 58, 1017–1025.

Vince, M.A. (1970) Some aspects of hatching behaviour. In: Freeman, B.M. and Gordon, R.F. (eds) *Aspects of Poultry Behaviour*. British Poultry Science Ltd, Edinburgh, UK, pp. 33–62.

Wabeck, C. (2002) Quality assurance and food safety – chicken meat. In: Bell, D.D. and Weaver, W.D. (eds) *Commercial Chicken Meat and Egg Production*, 5th edn. Kluwer, Dordrecht, The Netherlands, pp. 871–887.

Wahlström, A., Tauson, R. and Elwinger, K. (1998) Effects on plumage condition, health and mortality of dietary oats/wheat ratios to three hybrids of laying hens in different housing system. *Acta Agriculturae Scandinavica* 48, 250–259.

Walker, A.W., Alvey, D.M. and Tucker, S.A. (1997) Effect of an elevated food trough on bone strength and ease of catching laying hens. *British Poultry Science* 38, S14–S15.

Walker, P.M.B. (ed.) (1988) *Chambers Science and Technology Dictionary.* Chambers, Cambridge.

Wall, H. and Tauson, R. (2002) Egg quality in furnished cages for laying hens – effects of crack reduction methods and hybrid. *Poultry Science* 81, 340–348.

Wathes, C. and Charles, D. (eds) (1994) *Livestock Housing.* CAB International, Wallingford, UK.

Watts, C.R. and Stokes, A.W. (1971) The social order of turkeys. *Scientific American* 224, 112–118.

Webster, A.B. and Hurnik, J.F. (1994) Synchronization of behaviour among laying hens in battery cages. *Applied Animal Behaviour Science* 40, 153–165.

Wechsler, B. and Huber-Eicher, B. (1998) The effect of foraging material and perch height on feather pecking and feather damage in laying hens. *Applied Animal Behaviour Science* 58, 131–141.

Wechsler, B., Huber-Eicher, B. and Nash, D.R. (1998) Feather pecking in growers: a study with individually marked birds. *British Poultry Science* 39, 178–185.

Weeks, C.A., Nicol, C.J., Sherwin, C.M. and Kestin, S.C. (1994) Comparison of the behaviour of broiler chickens in indoor and free-range environments. *Animal Welfare* 3, 179–192.

Weeks, C.A., Danbury, B.D., Davies, H.C., Hunt, P. and Kestin, S.C. (2000) The behaviour of broiler chickens and its modification by lameness. *Applied Animal Behaviour Science* 67, 111–125.

Wegner, R.M. (1986) Alternative Systeme fur Legehennen – Untersuchungen in Europa. In: Larbier, M. (ed.) *Proceedings, 7th European Poultry Conference.* World's Poultry Science Association, Paris, pp. 1064–1076.

Wegner, R.M. (1990) Experience with the get-away cage system. *World's Poultry Science Journal* 46, 41–47.

Weigend, S. and Romanov, M.N. (2003) The world watch list for domestic animal diversity in the context of conservation and utilisation of poultry biodiversity. *World's Poultry Science Journal* 58, 411–430.

Wennrich, G. (1975) Studien zum Verhalten verschiedener Hybrid-Herkünfte von Haushühnern (*Gallus domesticus*) in Bodenintensivhaltung mit besonderer Berücksichtigung aggressiven Verhaltens sowie des Federpickens und des Kannibalismus. 5. Mitteilung: Verhaltensweisen des Federpickens. *Archiv für Geflügelkunde* 39, 37–44.

West, B. and Zhou, B.-X. (1989) Did chickens go north? New evidence for domestication. *World's Poultry Science Journal* 45, 205–218.

Whittow, G.C. (ed.) (2000) *Sturkie's Avian Physiology,* 5th edn. Academic Press, San Diego, California.

Widowski, T.M. and Duncan, I.J.H. (2000) Working for a dustbath: are hens increasing pleasure rather than reducing suffering? *Applied Animal Behaviour Science* 68, 39–53.

Widowski, T.M., Wong, D.M.A.L.F. and Duncan, I.J.H. (1998) Rearing with males accelerates onset of sexual maturity in female domestic fowl. *Poultry Science* 77, 150–155.

Wilson, B.W., Nieberg, P.S. and Buhr, R.J. (1990) Turkey muscle growth and focal myopathy. *Poultry Science* 59, 1553–1562.

Wilson, H.R., Piesco, N.P., Miller, E.R. and Nesbeth, W.G. (1979) Prediction of the fertility potential of broiler breeder males. *World's Poultry Science Journal* 35, 95–118.

Wolffram, R., Simons, J., Giebel, A. and Bongaerts, R. (2002) Impacts of stricter legal standards in the EU for keeping laying hens in battery cages. *World's Poultry Science Journal* 58, 365–370.

Woodard, A.E. and Abplanalp, H (1967) Semen production and fertilizing capacity of semen from Broad Breasted Bronze turkeys maintained in cages and on the floor. *Poultry Science* 46, 823–826.

Woodard, A.E. and Wilson, W.O. (1970) Behavioral patterns associated with oviposition in Japanese quail and chickens. *Journal of Interdisciplinary Cycle Research* 1, 173–180.

Wood-Gush, D.G.M. (1956) The agonistic and courtship behaviour of the brown Leghorn cock. *British Journal of Animal Behaviour* 4, 133–142.

Wood-Gush, D.G.M. (1959a) A history of the domestic fowl from antiquity to the 19th century. *Poultry Science* 38, 321–326.

Wood-Gush, D.G.M. (1959b) Time-lapse photography: a technique for studying diurnal rhythms. *Physiological Zoology* 32, 321–326.

Wood-Gush, D.G.M. (1960) A study of sex drive of two strains of cockerels through three generations. *Animal Behaviour* 8, 43–53.

Wood-Gush, D.G.M. (1963a) The relationship between hormonally-induced sexual behaviour in male chicks and their adult sexual behaviour. *Animal Behaviour* 11, 400–402.

Wood-Gush, D.G.M. (1963b) The control of nesting behaviour of the domestic hen. I. The role of the oviduct. *Animal Behaviour* 11, 293–299.

Wood-Gush, D.G.M. (1971) *The Behaviour of the Domestic Fowl.* Heinemann, London.

Wood-Gush, D.G.M. (1972) Strain differences in response to sub-optimal stimuli in the fowl. *Animal Behaviour* 20, 72–76.

Wood-Gush, D.G.M. and Gilbert, A.B. (1964) The control of the nesting behaviour of the domestic hen. II. The role of the ovary. *Animal Behaviour* 12, 451–453.

Wood-Gush, D.G.M. and Gilbert, A.B. (1970) The rate of egg loss through internal laying. *British Poultry Science* 11, 161–163.

Wood-Gush, D.G.M. and Gilbert, A.B. (1973) Some hormones involved in the nesting behaviour of hens. *Animal Behaviour* 21, 98–103.

Wood-Gush, D.G.M. and Gilbert, A.B. (1975) The physiological basis of a behaviour pattern in the domestic hen. *Symposia of the Zoological Society of London* 35, 261–276.

Wood-Gush, D.G.M. and Kare, M.R. (1966) The behaviour of calcium-deficient chickens. *British Poultry Science* 7, 285–290.

Wood-Gush, D.G.M. and Murphy, L.B. (1970) Some factors affecting the choice of nests by the hen. *British Poultry Science* 11, 415–417.

Wood-Gush, D.G.M. and Osborne, R. (1956) A study of differences in the sex drive of the domestic cock. *Animal Behaviour* 6, 68–71.

Wood-Gush, D.G.M., Duncan, I.J.H. and Savory, C.J. (1978) Observations on the social behaviour of domestic fowl in the wild. *Biology of Behaviour* 3, 193–205.

Wylie, L. (1999) Factors affecting poor breast feathering in modern turkeys. PhD Thesis, University of Edinburgh, Edinburgh, UK.

Yamada, Y. (1988) The contribution of poultry science to society. *World's Poultry Science Journal* 44, 172–178.

Yeomans, M.R. (1987) Control of drinking in domestic fowls. PhD Thesis, University of Edinburgh, Edinburgh, UK.

Ylander, D.M. and Craig, J.V. (1980) Inhibition of agonistic acts between domestic hens by a dominant third party. *Applied Animal Ethology* 6, 63–69.

Zeltner, E., Klein, T. and Huber-Eicher, B. (2000) Is there social transmission of feather pecking in groups of laying hen chicks? *Animal Behaviour* 60, 211–216.

Zeuner, E.E. (1963) *A History of Domesticated Animals.* Hutchinson, London.

Zuk, M., Thornhill, R., Ligon, J.D., Johnson, K., Austad, S., Ligon, S.H., Thornhill, N. and Costin, C. (1990) The role of male ornaments and courtship behavior in female choice of red junglefowl. *American Naturalist* 136, 459–473.

Zuk, M., Ligon, J.D. and Thornhill, R. (1992) Effects of experimental manipulation of male secondary sexual characteristics on female mate preference in red junglefowl. *Animal Behaviour* 44, 999–1006.

Zuk, M., Popma, S.L. and Johnsen, T.S. (1995) Male courtship displays, ornaments, and female mate choice in captive red junglefowl. *Behaviour* 132, 11–12.

Index

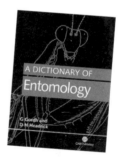
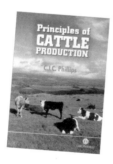